C-1414
ISBN 0-8373-1414-3

THE PASSBOOK® SERIES

PASSBOOKS®

FOR

CAREER OPPORTUNITIES

FOREMAN MACHINIST

NLC

National Learning Corporation
212 Michael Drive, Syosset, New York 11791
(516) 921-8888

National Learning Corporation

212 Michael Drive, Syosset, New York 11791

(516) 921-8888

PRINTED IN THE UNITED STATES OF AMERICA

PASSBOOK®

NOTICE

NATIONAL LEARNING CORPORATION
212 Michael Drive
Syosset, New York 11791

Inquiries re licensing agreements should be addressed to:
The President
National Learning Corporation
212 Michael Drive
Syosset, New York 11791

PASSBOOK SERIES®

The *PASSBOOK SERIES®* has been created to prepare applicants and candidates for the ultimate academic battlefield – the examination room.

At some time in our lives, each and every one of us may be required to take an examination – for validation, matriculation, admission, qualification, registration, certification, or licensure.

Based on the assumption that every applicant or candidate has met the basic formal educational standards, has taken the required number of courses, and read the necessary texts, the *PASSBOOK SERIES®* furnishes the one special preparation which may assure passing with confidence, instead of failing with insecurity. Examination questions – together with answers – are furnished as the basic vehicle for study so that the mysteries of the examination and its compounding difficulties may be eliminated or diminished by a sure method.

This book is meant to help you pass your examination provided that you qualify and are serious in your objective.

The entire field is reviewed through the huge store of content information which is succinctly presented through a provocative and challenging approach – the question-and-answer method.

A climate of success is established by furnishing the correct answers at the end of each test.

You soon learn to recognize types of questions, forms of questions, and patterns of questioning. You may even begin to anticipate expected outcomes.

You perceive that many questions are repeated or adapted so that you gain acute insights, which may enable you to score many sure points.

You learn how to confront new questions, or types of questions, and to attack them confidently and work out the correct answers.

You note objectives and emphases, and recognize pitfalls and dangers, so that you may make positive educational adjustments.

Moreover, you are kept fully informed in relation to new concepts, methods, practices, and directions in the field.

You discover that you are actually taking the examination all the time: you are preparing for the examination by "taking" an examination, not by reading extraneous and/or supererogatory textbooks.

In short, this PASSBOOK®, used directedly, should be an important factor in helping you to pass your test.

FOREMAN MACHINIST

DUTIES AND RESPONSIBILITIES

Under general direction, supervises, directs, and is responsible for the operation of a machine shop in connection with machine and bench work normally performed therein. Performs related work.

EXAMPLES OF TYPICAL TASKS

Schedules and is responsible for the work of Machinists and/or Auto Machinists and other assigned personnel in altering, repairing, and fabricating tools, machine components, and marine machinery in shop and field. Interprets drawings and makes decisions relevant to work methods. Prepares work schedules and makes work assignments. Uses instruments and gauges in checking work dimensions. Makes inspections and checks work programs. Is responsible for the maintenance and repair of bench and machine equipment common to a machine shop. Prepares job orders and requisitions for tools, materials, and equipment. Keeps records and makes reports.

TESTS

The written test may include questions on supervisory and administrative procedures, including work assigning and coordinating, discipline, motivation, training, human relations, and understanding of printed matter; writing reports, forms, ordering of materials, record keeping, safety methods; the machinist trade, machinery installation, and related tools; ability to translate administrative instructions into detailed operational plans for machine parts fabrications, repair, and installation, for both machine and automotive equipment; estimating time, costs, and material; blueprint or plan reading, basic mathematical computations; and other related areas.

HOW TO TAKE A TEST

I. YOU MUST PASS AN EXAMINATION

A. WHAT EVERY CANDIDATE SHOULD KNOW

Examination applicants often ask us for help in preparing for the written test. What can I study in advance? What kinds of questions will be asked? How will the test be given? How will the papers be graded?

As an applicant for a civil service examination, you may be wondering about some of these things. Our purpose here is to suggest effective methods of advance study and to describe civil service examinations.

Your chances for success on this examination can be increased if you know how to prepare. Those "pre-examination jitters" can be reduced if you know what to expect. You can even experience an adventure in good citizenship if you know why civil service examinations are given.

B. WHY ARE CIVIL SERVICE EXAMINATIONS GIVEN?

Civil service examinations are important to you in two ways. As a citizen, you want public jobs filled by employees who know how to do their work. As a job-seeker, you want a fair chance to compete for that job on an equal footing with other candidates. The best known means of accomplishing this two-fold goal is the competitive examination.

Examinations are widely publicized throughout the nation. They may be administered for jobs in federal, state, city, municipal, town, or village governments or agencies.

Any citizen may apply, with some limitations, such as the age or residence of applicants. Your experience and education may be reviewed to see whether you meet the requirements for the particular examination. When these requirements exist, they are reasonable and are applied consistently to all applicants. Thus, a competitive examination may cause you some uneasiness now, but it is your privilege and safeguard.

C. HOW ARE CIVIL SERVICE EXAMINATIONS DEVELOPED?

Examinations are carefully written by trained technicians who are specialists in the field known as "psychological measurement," in consultation with recognized authorities in the field of work that the test will cover. These experts recommend the subject matter areas or skills to be tested; only those knowledges or skills important to your success on the job are included. The most reliable books and source materials available are used as references. Together, the experts and technicians judge the difficulty level of the questions.

Test technicians know how to phrase questions so that the problem is clearly stated. Their ethics do not permit "trick" or "catch" questions. Questions may have been tried out on sample groups, or subjected to statistical analysis, to determine their usefulness.

Written tests are often used in combination with performance tests, ratings of training and experience, and oral interviews. All of these measures combine to form the best known means of finding the right man for the right job.

II. HOW TO PASS THE WRITTEN TEST

A. NATURE OF THE EXAMINATION

To prepare intelligently for civil service examinations, you should know how they differ from school examinations you have taken. In school you were assigned certain definite pages to read or subjects to cover. The examination questions were quite detailed and usually emphasized memory. Civil service examinations, on the other hand, try to discover your present ability to perform the duties of a position, plus your potentiality to learn these duties. In other words, a civil service examination attempts to predict how successful you will be. Questions cover such a broad area that they cannot be as minute and detailed as school examination questions.

In the public service similar kinds of work, or positions, are grouped together in one "class." This process is known as "position-classification." All the positions in a class are paid according to the salary range for that class. One class title covers all these positions, and they are all tested by the same examination.

B. FOUR BASIC STEPS

1. Study the Announcement.--How, then, can you know what subjects to study? Our best answer is: "Learn as much as possible about the class of positions for which you have applied." The examination will test the knowledge, skills, and abilities needed to do the work.

Your most valuable source of information about the position you want is the official announcement of the examination. This announcement lists the training and experience qualifications. Check these standards and apply only if you come reasonably close to meeting them.

The brief description of the position in the examination announcement offers some clues to the subjects which will be tested. Think about the job itself. Review the duties in your mind. Can you perform them, or are there some in which you are rusty? Fill in the blank spots in your preparation.

Many jurisdictions preview the written test in the examination announcement by including a section called "Knowledge and Abilities Required," "Scope of Examination," or some similar heading. Here you will find out specifically what fields will be tested.

2. Review Your Own Background.-- Once you learn in general what the position is all about, and what you need to know to do the work, ask yourself which subjects you already know fairly well and which need improvement. You may wonder whether to concentrate on improving your strong areas or on building some background in your fields of weakness. When the announcement has specified "some knowledge" or "considerable knowledge," or has used adjectives such as "beginning principles of ...." or "advanced .....methods," you can get a clue as to the number and difficulty of questions to be asked in any given field. More questions, and hence broader coverage, would be included for those subjects which are more important in the work. Now weigh your strengths and weaknesses against the job requirements and prepare accordingly.

3. Determine the Level of the Position.-- Another way to tell how intensively you should prepare is to understand the level of the job for which you are applying. Is it the entering level? In other words, is this the position in which beginners in a field of work are hired? Or is it an intermediate or advanced level? Sometimes this is indicated by such words as "Junior" or "Senior" in the class title.Other jurisdictions use Roman numerals to designate the level: Clerk I,

Clerk II, for example. The word "Supervisor" sometimes appears in the title. If the level is not indicated by the title, check the description of duties. Will you be working under very close supervision, or will you have responsibility for independent decisions in this work?

4. Choose Appropriate Study Materials.-- Now that you know the subjects to be examined and the relative amount of each subject to be covered, you can choose suitable study materials. For beginning level jobs, or even advanced ones, if you have a pronounced weakness in some aspect of your training, read a modern, standard textbook in that field. Be sure it is up-to-date and has general coverage. Such books are normally available at your library, and the librarian will be glad to help you locate one. For entry level positions, questions of appropriate difficulty are chosen -- neither highly advanced questions, nor those too simple. Such questions require careful thought but not advanced training.

If the position for which you are applying is technical or advanced, you will read more advanced, specialized material. If you are already familiar with the basic principles of your field, elementary textbooks would waste your time. Concentrate on advanced textbooks and technical periodicals. Think through the concepts and review difficult problems in your field.

These are all general sources. You can get more ideas on your own initiative, following these leads. For example, training manuals and publications of the government agency which employs workers in your field can be useful, particularly for technical and professional positions. A letter or visit to the government department involved may result in more specific study suggestions, and certainly will provide you with a more definite idea of the exact nature of the position you are seeking.

III. KINDS OF TESTS

Tests are used for purposes other than measuring knowledge and ability to perform specified duties. For some positions, it is equally important to test ability to make adjustments to new situations or to profit from training. In others, basic mental abilities not dependent upon information are essential. Questions which test these things may not appear as pertinent to the duties of the position as those which test for knowledge and information. Yet they are often highly important parts of a fair examination. For very general questions, it is almost impossible to help you direct your study efforts. What we can do is to point out some of the more common of these general abilities needed in public service positions and describe some typical questions.

1. General Information

Broad, general information has been found useful for predicting job success in some kinds of work. This is tested in a variety of ways, from vocabulary lists to questions about current events. Basic background in some field of work, such as sociology or economics, may be sampled in a group of questions. Often these are principles which have become familiar to most persons through "exposure" rather than through formal training. It is difficult to advise you how to study for these questions; being alert to the world around you is our best suggestion.

2. Verbal Ability

An example of an ability needed in many positions is verbal or language ability. Verbal ability is, in brief, the ability to use and understand words. Vocabulary and grammar tests are typical measures of this ability. "Reading comprehension" or "paragraph interpretation" questions are common in many kinds of civil servic tests. You are given a paragraph of written material and asked to find its central meaning.

3. Numerical Ability

Number skills can be tested by the familiar arithmetic problem, by checking paired lists of numbers to see which are alike an which are different, or by interpreting charts and graphs. In the latter test, a graph may be printed in the test booklet which you are asked to use as the basis for answering questions.

4. Observation

A popular test for law-enforcement positions is the observation test. A picture is shown to you for several minutes, then taken away. Questions about the picture test your ability to observe both details and larger elements.

5. Following Directions

In many positions in the public service, the employee must be able to carry out written instructions dependably and accurately. You may be given a chart with several columns, each column lis ing a variety of information. The questions require you to carry o directions involving the information given in the chart.

6. Skills and Aptitudes

Performance tests effectively measure some manual skills an aptitudes. When the skill is one in which you are trained, such as typing or shorthand, you can practice. These tests are often very much like those given in business school or high school courses. For many of the other skills and aptitudes, however, no short-time preparation can be made. Skills and abilities natural to you or that you have developed throughout your lifetime are being tested.

Many of the general questions just described provide all the data needed to answer the questions and ask you to use your reasoning ability to find the answers. Your best preparation for these tests as well as for tests of facts and ideas, is to be at your physical and mental best. You, no doubt, have your own methods of getting into an exam-taking mood and keeping "in shape." The next section lists some ideas on this subject.

IV. KINDS OF QUESTIONS

Only rarely is the "essay" question, which you answer in narrative form, used in civil service tests. Civil service tests are usually of the short-answer type. Full instructions for answering these questions will be given to you at the examination. But in case this is your first experience with short-answer questions and separate answer sheets, here is what you need to know.

1. Multiple-Choice Questions

Most popular of the short-answer questions is the "multiple-choice" or "best-answer" question. It can be used, for example, to test for factual knowledge, ability to solve problems, or judgment in meeting situations found at work.

A multiple-choice question is normally one of three types:

(1) It can begin with an incomplete statement followed by several possible endings. You are to find the one ending which *best* completes the statement, although some of the others may not be entirely wrong.

(2) It can also be a complete statement in the form of a question which is answered by choosing one of the statements listed.

(3) It can be in the form of a problem -- again you select the best answer.

Here is an example of a multiple-choice question with a discussion which should give you some clues as to the method for choosing the right answer.

SAMPLE QUESTION:

When an employee has a complaint about his assignment, the action which will *best* help him overcome his difficulty is

(A) to discuss his difficulty with his co-workers
(B) to take the problem to the head of the organization
(C) to take the problem to the person who gave him the assignment
(D) to say nothing to anyone about his complaint

In answering this question you should study each of the choices to find which is best. Consider choice (A). Certainly an employee may discuss his complaint with fellow employees, but no change or improvement can result, and the complaint remains unsolved. Choice (B) is a poor choice since the head of the organization probably does not know what assignment you have been given, and taking your problem to him is known as "going over the head" of the supervisor. The supervisor, or person who made the assignment, is the person who can clarify it or correct any injustice. Choice (C) is, therefore, correct. To say nothing, as in choice (D), is unwise. Supervisors have an interest in knowing the problems employees are facing, and the employee is seeking a solution to his problem.

2. True-False Questions

The "true-false" or "right-wrong" form of question is sometimes used. Here a complete statement is given. Your problem is to decide whether the statement is right or wrong.

SAMPLE QUESTION:

A person-to-person long distance telephone call costs less than a station-to-station call to the same city.

This question is wrong, or "false," since person-to-person calls are more expensive.

This is not a complete list of all possible question forms, although most of the others are variations of these common types. You will always get complete directions for answering questions. Be sure you understand *how* to mark your answers -- ask questions until you do.

V. RECORDING YOUR ANSWERS

For an examination with very few applicants, you may be told to record your answers in the test booklet itself. Separate answer sheets are much more common. If this separate answer sheet is to be scored by machine -- and this is often the case -- it is highly important that you mark your answers correctly in order to get credit.

An electric test-scoring machine is often used in civil service offices because of the speed with which papers can be scored. Machine-scored answer sheets must be marked with a special pencil, which will be given to you. This pencil has a high graphite conten which responds to the electrical scoring machine. As a matter of fact, stray dots may register as answers, so do not let your pencil rest on the answer sheet while you are pondering the correct answer Also, if your pencil lead breaks or is otherwise defective, ask for another.

Since the answer sheet will be dropped in a slot in the scoring machine, be careful not to bend the corners or get the paper crumpled.

The answer sheet normally has five vertical columns of numbers, with 30 numbers to a column. These numbers correspond to the question numbers in your test booklet. After each number, going across the page, are four or five pairs of dotted lines. These short dotted lines have small letters or numbers above them. The first two pairs may also have a "T" and "F" above the letters. This indicates that the first two pairs only are to be used if the questions are of the true-false type. If the questions are multiple-choice, disregard this "T" and "F" completely, and pay attention only to the small number or letters.

Answer your questions in the manner of the sample that follows. Proceed in the sequential steps outlined below.

Assume that you are answering question 32, which is:

32. The largest city in the United States is:

A. Washington, D.C. B. New York City C. Chicago
D. Detroit E. San Francisco

1. Choose the answer you think is best.
   New York City is the largest, so choice B is correct.
2. Find the row of dotted lines numbered the same as the question you are answering.
   This is question number 32, so find row number 32.
3. Find the pair of dotted lines corresponding to the answer you have chosen.
   You have chosen answer B, so find the pair of dotted lines marked "B".
4. Make a solid black mark between the dotted lines.
   Go up and down two or three times with your pencil so plenty of graphite rubs off, but do not let the mark get outside or above the dots.

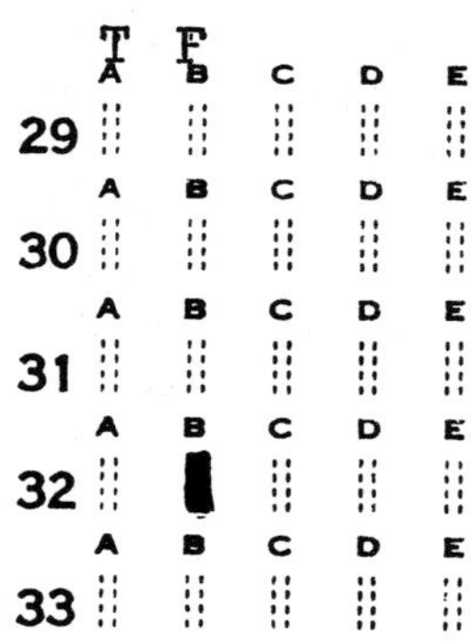

VI. BEFORE THE TEST

Common sense will help you find procedures to follow to get ready for an examination. Too many of us, however, overlook these sensibl measures. Indeed, nervousness and fatigue have been found to be the most serious reasons why applicants fail to do their best on civil service tests. Here is a list of reminders.

1. Begin Your Preparation Early

Don't wait until the last minute to go scurrying around for books and materials or to find out what the position is all about.

2. Prepare Continuously

An hour a night for a week is better than an all-night cram session. This has been definitely established. What is more, a night a week for a month will return better dividends than crowding your study into a shorter period of time.

3. Locate the Place of the Examination

You have been sent a notice telling you when and where to report for the examination. If the location is in a different town or otherwise unfamiliar to you, it would be well to inquire the best route and learn something about the building.

4. Relax the Night Before the Test

Allow your mind to rest. Do not study at all that night. Plan some mild recreation or diversion; then go to bed early and get a good night's sleep.

5. Get Up Early Enough to Make a Leisurely Trip to the Place for the Test

Then unforeseen events, traffic snarls, unfamiliar buildings, will not upset you.

6. Dress Comfortably

A written test is not a fashion show. You will be known by number and not by name, so wear something comfortable.

7. Leave Excess Paraphernalia at Home

Shopping bags and odd bundles will get in your way. You need bring only the items mentioned in the official notice sent to you; usually everything you need is provided. Do not bring reference books to the examination. They will only confuse those last minutes and be taken away from you when in the test room.

8. Arrive Somewhat Ahead of Time

If because of transportation schedules you must get there very early, bring a newspaper or magazine to take your mind off yourself while waiting.

9. Locate the Examination Room

When you have found the proper room, you will be directed to the seat or part of the room where you will sit. Sometimes you are given a sheet of instructions to read while you are waiting. Do not fill out any forms until you are told to do so; just read them and be ready.

10. Relax and Prepare to Listen to the Instructions

11. If you have any physical problem that may keep you from doing your best, be sure to tell the test administrator. If you are sick, or in poor health, you really cannot do your best on the test. You can come back and take the test some other time.

VII. AT THE TEST

The day of the test is here and you have the test booklet in your hand. The temptation to get going is very strong. Caution! There is more to success than knowing the right answers. You must know how to identify your papers and understand variations in the type of short-answer question used in this particular examination. Follow these suggestions for maximum results from your efforts.

1. Cooperate with the Monitor

The test administrator has a duty to create a situation in which you can be as much at ease as possible. He will give instructions, tell you when to begin, check to see that you are marking your answer sheet correctly. He is not there to guard you, although he will see that your competitors do not take unfair advantage. He wants to help you do your best.

2. Listen to All Instructions

Don't jump the gun! Wait until you understand all directions. In most civil service tests you get more time than you need to answer the questions. So don't get in a hurry. Read each word of instructions until you clearly understand the meaning. Study the examples. Listen to all announcements. Follow directions. Ask questions if you do not understand what to do.

3. Identify Your Papers

Civil service examinations are usually identified by number only. You will be assigned a number; you must not put your name on your test papers. Be sure to copy your number correctly. Since more than one examination may be given, copy your exact examination title.

4. Plan Your Time

Unless you are told that a test is a "speed" or "rate-of-work" test, speed itself is not usually important. Time enough to answer all the questions will be provided. But this does not mean that you have all day. An overall time limit has been set. Divide the total time (in minutes) by the number of questions to get the approximate time you have for each question.

5. Do Not Linger Over Difficult Questions

If you come across a difficult question, mark it with a paper clip (useful to have along) and come back to it when you have been through the booklet. One caution if you do this -- be sure to skip a number on your answer sheet too. Check often to be sure that you have not lost your place and that you are marking in the row numbered the same as the question you are answering.

6. Read the Questions

Be sure you know what the question asks! Many capable people are unsuccessful because they failed to *read* the questions correctly.

7. Answer All Questions

Unless you have been instructed that a penalty will be deducted for incorrect answers, it is better to guess than to omit a question.

8. Speed Tests

It is often better *not* to guess on speed tests. It has been found that on timed tests people are tempted to spend the last few seconds before time is called in marking answers at random -- without even reading them -- in the hope of picking up a few extra points. To discourage this practice, the instructions may warn you that your score will be "corrected" for guessing. That is, a penalty will be applied. The incorrect answers will be deducted from the correct ones, or some other penalty formula will be used.

9. Review Your Answers

If you finish before time is called, go back to the questions you guessed or omitted to give further thought to them. Review other answers if you have time.

10. Return Your Test Materials

If you are ready to leave before others have finished or time is called, take *all* your materials to the monitor and leave quietly. Never take any test material with you. The monitor can discover whose papers are not complete, and taking a test booklet may be grounds for disqualification.

II. EXAMINATION TECHNIQUES

1. Read the *general* instructions carefully. These are usually printed on the first page of the examination booklet. As a rule, these instructions refer to the timing of the examination; the fact that you should not start work until the signal and must stop work at a signal, etc. If there are any *special* instructions, such as a choice of questions to be answered, make sure that you note this instruction carefully.

2. When you are ready to start work on the examination, that is as soon as the signal has been given, read the instructions to each question booklet, underline any key words or phrases, such as *least, best, outline, describe,* and the like. In this way you will tend to answer as requested rather than discover on reviewing your paper that you *listed without describing,* that you selected the *worst* choice rather than the *best* choice, etc.

3. If the examination is of the objective or so-called multiple-choice type, that is, each question will also give a series of possible answers: A, B, C, or D, and you are called upon to select the best answer and write the letter next to that answer on your answer paper, it is advisable to start answering each question in turn. There may be anywhere from 50 to 100 such questions in the three or four hours allotted and you can see how much time would be taken if you read through all the questions before beginning to answer any. Furthermore, if you come across a question or a group of questions which you know would be difficult to answer, it would undoubtedly affect your handling of all the other questions.

4. If the examination is of the esssay-type and contains but a few questions, it is a moot point as to whether you should read all the questions before starting to answer any one. Of course if you are given a choice, say five out of seven and the like, then it is essential to read all the questions so you can eliminate the two which are most difficult. If, however, you are asked to answer all the questions, there may be danger in trying to answer the easiest one first because you may find that you will spend too much time on it. The best technique is to answer the first question, then proceed to the second, etc.

5. Time your answers. Before the examination begins, write down the time it started, then add the time allowed for the examination and write down the time it must be completed, then divide the time available somewhat as follows:

(a) If 3½ hours are allowed, that would be 210 minutes. If you have 80 objective-type questions, that would be an average of 2½ minutes per question. Allow yourself no more than 2 minutes per question, or a total of 160 minutes, which will permit about 50 minutes to review.

(b) If for the time allotment of 210 minutes, there are 7 essay questions to answer, that would average about 30 minutes a question. Give yourself only 25 minutes per question so that you have about 35 minutes to review.

6. The most important instruction is *to read each question* and make sure you know what is wanted. The second most important instruction is to *time yourself properly* so that you answer every question. The third most important instruction is to *answer every question.* Guess if you have to but include something for each question. Remember that you will receive no credit for a blank and will probably receive some credit if you write something in answer to an essay question. If you guess a letter, say "B" for a multiple-choice question, you may have guessed right. If you leave a blank as the answer to a multiple-choice question, the examiners may respect your feelings but it will not add a point to your score.

7. Suggestions

a. Objective-Type Questions

(1) Examine the question booklet for proper sequence of pages and questions.
(2) Read all instructions carefully.
(3) Skip any question which seems too difficult; return to it after all other questions have been answered.
(4) Apportion your time properly; do not spend too much time on any single question or group of questions.
(5) Note and underline key words -- *all, most, fewest, least, best, worst, same, opposite.*
(6) Pay particular attention to negatives.
(7) Note unusual option, e.g., unduly long, short, complex, different or similar in content to the body of the question.
(8) Observe the use of "hedging" words -- *probably, may, most likely, etc.*
(9) Make sure that your answer is put next to the same number as the question.
(10) Do not second-guess unless you have good reason to believe the second answer is definitely more correct.
(11) Cross out original answer if you decide another answer is more accurate; do not erase.
(12) Answer all questions; guess unless instructed otherwise.
(13) Leave time for review.

b. Essay-Type Questions

(1) Read each question carefully.
(2) Determine exactly what is wanted. Underline key words or phrases.
(3) Decide on outline or paragraph answer.
(4) Include many different points and elements unless asked to develop any one or two points or elements.
(5) Show impartiality by giving pros and cons unless directed to select one side only.
(6) Make and write down any assumptions you find necessary to answer the question.
(7) Watch your English, grammar, punctuation, choice of words.
(8) Time your answers; don't crowd material.

8. Answering the Essay Question

Most essay questions can be answered by framing the specific response around several key words or ideas. Here are a few such key words or ideas:

M's: manpower, materials, methods, money, management;
P's: purpose, program, policy, plan, procedure, practice, problems, pitfalls, personnel, public relations.

a. Six Basic Steps in Handling Problems:
(1) Preliminary plan and background development
(2) Collect information, data and facts
(3) Analyze and interpret information, data and facts
(4) Analyze and develop solutions as well as make recommendations
(5) Prepare report and sell recommendations
(6) Install recommendations and follow up effectiveness

b. Pitfalls to Avoid
(1) *Taking things for granted*
A statement of the situation does not necessarily imply that each of the elements is necessarily true; for example, a complaint may be invalid and biased so that all that can be taken for granted is that a complaint has been registered.
(2) *Considering only one side of a situation*
Wherever possible, indicate several alternatives and then point out the reasons you selected the best one.
(3) *Failing to indicate follow-up*
Whenever your answer indicates action on your part, make certain that you will take proper follow-up action to see how successful your recommendations, procedures, or actions turn out to be.
(4) *Taking too long in answering any single question*
Remember to time your answers properly.

IX. AFTER THE TEST

Scoring procedures differ in detail among civil service jurisdictions although the general principles are the same. Whether the papers are hand-scored or graded by the electric scoring machine we have described, they are nearly always graded by number. That is, the person who marks the paper knows only the number -- never the name -- of the applicant. Not until all the papers have been graded will they be matched with names. If other tests, such as training and experience or oral interview ratings have been given, scores will be combined. Different parts of the examination usually have different weights. For example, the written test might count 60 percent of the final grade, and a rating of training and experience 40 percent. In many jurisdictions, veterans will have a certain number of points added to their grades.

After the final grade has been determined, the names are placed in grade order and an eligible list is established. There are various methods for resolving ties between those who get the same final grade: probably the most common is to place first the name of the person whose application was received first. Job offers are made from the eligible list in the order the names appear on it.

You will be notified of your grade and your rank order as soon as all these computations have been made. This will be done as rapidly as possible.

People who are found to meet the requirements in the announcement are called "eligibles." Their names are put on a list of eligibles. An eligible's chances of getting a job depend on how high he stands on this list and how fast agencies are filling jobs from the list.

When a job is to be filled from a list of eligibles, the agency asks for the names of people on the list of eligibles for that job.

When the civil service commission receives this request, it sends to the agency the names of the three people highest on the list. Or, if the job to be filled has specialized requirements, the office sends the agency, from the general list, the names of the top three persons who meet those requirements.

The appointing officer makes a choice from among the three people whose names were sent to him. If the selected person accepts the appointment, the names of the others are put back on the list to be considered for future openings.

That is the rule in hiring from all kinds of eligible lists, whether they are for typist, carpenter, chemist, or something else. For every vacancy, the appointing officer has his choice of any one of the top three eligibles on the list. This explains why the person whose name is on top of the list sometimes does not get an appointment when some of the persons lower on the list do. If the appointing officer chooses the No.2 or No.3 eligible, the No.1 eligible does not get a job at once, but stays on the list until he is appointed or the list is terminated.

X. HOW TO PASS THE INTERVIEW TEST

The examination for which you applied requires an oral interview test. You have already taken the written test and you are now being called for the interview test -- the final part of the formal examination.

You may think that it is not possible to prepare for an interview test and that there are no procedures to follow during an interview. Our purpose is to point out some things you can do in advance that will help you and some good rules to follow and pitfalls to avoid while you are being interviewed.

A. WHAT IS AN INTERVIEW SUPPOSED TO TEST?

The written examination is designed to test the technical knowledge and competence of the candidate; the oral is designed to evaluate intangible qualities, not readily measured otherwise, and to establish a list showing the relative fitness of each candidate, *as measured against his competitors*, for the position sought. Scoring is not on the basis of "right" or "wrong," but on a sliding scale of values ranging from "not passable" to "outstanding." As a matter of fact, it is possible to achieve a relatively low score without a single "incorrect" answer because of evident weakness in the qualities being measured,

Occasionally, an examination may consist entirely of an oral test -- either an individual or a group oral. In such cases, information is sought concerning the technical knowledges and abilities of the candidate, since there has been no written examination for this purpose. More commonly, however, an oral test is used to supplement a written examination.

B. WHO CONDUCTS INTERVIEWS?

The composition of oral boards varies among different jurisdictions. In nearly all, a representative of the personnel department serves as chairman. One of the members of the board may be a representative of the department in which the candidate would work. In some cases, "outside experts" are used, and, frequently, a business man or some other representative of the general public is asked to serve. Labor and management or other special groups may be represented. The aim is to secure the services of experts in the appropriate field.

However the board is composed, it is a good idea (and not at all improper or unethical) to ascertain in advance of the interview who the members are and what groups they represent. When you are introduced to them, you will have some idea of their backgrounds and interests, and at least you will not stutter and stammer over their names.

C. WHAT TO DO BEFORE THE INTERVIEW

While knowledge about the board members is useful and takes some of the surprise element out of the interview, there is other preparation which is more substantive. It *is* possible to prepare for an oral -- in several ways:

1. <u>Keep a Copy of Your Application and Review it Carefully Before the Interview</u>

   This may be the only document before the oral board, and the starting point of the interview. Know what experience and education you have listed there, and the sequence and dates of it. Sometimes the board will ask *you* to review the highlights of your experience for them; you should not have to hem and haw doing it.

2. <u>Study the Class Specification and the Examination Announcement</u>

   Usually, the oral board has one or both of these to guide them. The qualities, characteristics, or knowledges required by the position sought are stated in these documents. They offer valuable clues as to the nature of the oral interview. For example, if the job involves supervisory responsibilities, the announcement will usually indicate that knowledge of modern supervisory methods and the qualifications of the candidate as a supervisor will be tested. If so, you can expect such questions, frequently in the form of a hypothetical situation which you are expected to solve. *Never* go into an oral without knowledge of the duties and responsibilities of the job you seek.

3. <u>Think Through Each Qualification Required</u>

   Try to visualize the kind of questions *you* would ask if you were a board member. How well could you answer them? Try especially to appraise your own knowledge and background in each area, *measured against the job sought*, and identify any areas in which you are weak. Be critical and realistic -- do not flatter yourself.

4. <u>Do Some General Reading in Areas in Which You Feel You May be Weak</u>

   For example, if the job involves supervision and your past experience has *not*, some general reading in supervisory methods and practices, particularly in the field of human relations, might be useful. *Do not* study agency procedures or detailed manuals. The oral board will be testing your understanding and capacity, *not* your memory.

5. <u>Get a Good Night's Sleep and Watch Your General Health and Mental Attitude</u>

   You will want a clear head at the interview. Take care of a cold or other minor ailment, and, of course, *no hangovers*.

D. WHAT TO DO THE DAY OF THE INTERVIEW

Now comes the day of the interview itself. Give yourself plenty of time to get there. Plan to arrive somewhat ahead of the scheduled time, particularly if your appointment is in the fore part of the day. If a previous candidate fails to appear, the board might be ready for you a bit early. By early afternoon an oral board is almost invariably behind schedule if there are many candidates, and you may have to wait. Take along a book or magazine to read, or your application to review. But leave any extraneous material in the waiting room when you go in for your interview. In any event, relax and compose yourself.

The matter of dress is important. The board is forming impressions about you -- from your experience, your manners, your attitudes, and from your appearance. Give your personal appearance careful attention. Dress your *best*, but not your flashiest. Choose conservative, appropriate clothing, and be sure it and you are immaculate. This is a business interview, and your appearance should indicate that you regard it as such. Besides, being well-groomed and properly dressed will help boost your confidence.

Sooner or later, someone will call your name and escort you into the interview room. *This is it.* From here on you are on your own. It is too late for any more preparation. But, remember, you asked for this opportunity to prove your fitness, and you are here because your request was granted.

E. WHAT HAPPENS WHEN YOU GO IN?

The usual sequence of events will be as follows: The clerk (who is often the board stenographer) will introduce you to the chairman of the oral board, who will introduce you to each other member of the board. Acknowledge the introductions before you sit down. Do not be surprised if you find a microphone facing you or a stenotypist sitting by. Oral interviews are usually recorded, in the event of an appeal or other review.

Usually the chairman of the board will open the interview by reviewing the highlights of your education and work experience from your application -- primarily for the benefit of the other members of the board, as well as to get the material into the record. Do not interrupt or comment unless there is an error or significant misinterpretation; if so, do not hesitate. But do not quibble about insignificant matters. Usually, also, he will ask you some question about your education, your experience, or your present job -- partly to get you started talking, to establish the interviewing "rapport." He may start the actual questioning, or turn it over to one of the other members. Frequently each member undertakes the questioning on a particular area, one in which he is perhaps most competent. So you can expect each member to participate in the examination. And because the time is limited, you may expect some rather abrupt switches in the direction the questioning takes. Do not be upset by it. Normally, a board member will not pursue a single line of questioning unless he discovers a particular strength or weakness.

After each member has participated, the chairman will usually ask whether any member has any further questions, then will ask you if you have anything you wish to add. Unless you are expecting this question, it may floor you. Or worse, it may start you off on an extended, extemporaneous speech. The board is not usually seeking more information. The question is principally to offer you a last opportunity to present further qualifications or to indicate that you have nothing to add. So, if you feel that a significant qualification or characteristic has been overlooked, it is proper to point it out in a sentence or so. Do not compliment the board on the thoroughness of their examination -- they have been sketchy, and you know it. If you wish, merely say, "No thank you, I have nothing further to add." This is a point where you can "talk yourself out" of a good impression or fail to present an important bit of information. *Remember, you close the interview yourself.*

The chairman will then say, "That is all, Mr. Smith, thank you." Do not be startled; the interview is over, and quicker than you think. Say, "Thank you and good morning," gather up your belongings and take your leave. Save your sigh of relief for the other side of the door.

F. HOW TO PUT YOUR BEST FOOT FORWARD

Throughout all this process, you may feel that the board individually and collectively is trying to pierce your defenses, to seek out your hidden weaknesses, and to embarrass and confuse you. Actually, this is not true. They are obliged to make an appraisal of your qualifications for the job you are seeking, and they *want to see you in your best light*. Remember, they must interview all candidates and a noncooperative candidate may become a failure in spite of their best efforts to bring out his qualifications. Here are fifteen(15) suggestions that will help you:

1. Be Natural. Keep Your Attitude Confident, But Not Cocky

If *you* are not confident that you can do the job, do not exexpect the *board* to be. Do not apologize for your weaknesses, try to bring out your strong points. The board is interested in a positive, not a negative presentation. Cockiness will antagonize any board member, and make him wonder if you are covering up a weakness by a false show of strength.

2. Get Comfortable, But Don't Lounge or Sprawl

Sit erectly but not stiffly. A careless posture may lead the board to conclude you are careless in other things, or at least that you are not impressed by the importance of the occasion to you. Either conclusion is natural, even if incorrect. Do not fuss with your clothing, or with a pencil or an ashtray. Your hands may occasionally be useful to emphasize a point; do not let them become a point of distraction.

3. Do Not Wisecrack or Make Small Talk

This is a serious situation, and your attitude should show that you consider it as such. Further, the time of the board is limited; they do not want to waste it, and neither should you.

4. Do Not Exaggerate Your Experience or Abilities

In the first place, from information in the application, from other interviews and other sources, the board may know more about you than you think; in the second place, you probably will not get away with it in the first place. An experienced board is rather adept at spotting such a situation. Do not take the chance.

5. If You Know a Member of the Board, Do Not Make a Point of It, Yet Do Not Hide It

Certainly you are not fooling him, and probably not the other members of the board. Do not try to take advantage of your acquaintanceship -- it will probably do you little good.

6. Do Not Dominate the Interview

Let the board do that. They will give you the clues -- do not assume that you have to do all the talking. Realize that the board has a number of questions to ask you, and do not try to take up all the interview time by showing off your extensive knowledge of the answer to the first one.

7. Be Attentive

You only have twenty minutes or so, and you should keep your attention at its sharpest throughout. When a member is addressing a problem or a question to you, give him your undivided attention. Address your reply principally to him, but do not exclude the other members of the board.

8. Do Not Interrupt

A board member may be stating a problem for you to analyze. He will ask you a question when the time comes. Let him state the problem, and wait for the question.

9. Make Sure You Understand the Question

Do not try to answer until you are sure what the question is. If it is not clear,restate it in your own words or ask the board member to clarify it for you. But do not haggle about minor elements.

10. Reply Promptly But Not Hastily

A common entry on oral board rating sheets is "candidate responded readily," or "candidate hesitated in replies." Respond as promptly and quickly as you can, but do not jump to a hasty, ill-considered answer.

11. Do Not Be Peremptory in Your Answers

A brief answer is proper -- but do not fire your answer back. That is a losing game from your point of view. The board member can probably ask questions much faster than you can answer them.

12. Do Not Try To Create the Answer You Think the Board Member Wants

He is interested in what kind of mind you have and how it works -- not in playing games. Furthermore, he can usually spot this practice and will usually grade you down on it.

13. Do Not Switch Sides in Your Reply Merely to Agree With a Board Member

Frequently, a member will take a contrary position merely to draw you out and to see if you are willing and able to defend your point of view. Do not start a debate, yet do not surrender a good position. If a position is worth taking, it is worth defending.

14. Do Not Be Afraid to Admit an Error in Judgment if You Are Shown to Be Wrong

The board knows that you are forced to reply without any opportunity for careful consideration. Your answer may be demonstrably wrong. If so, admit it and get on with the interview.

15. Do Not Dwell at Length on Your Present Job

The opening question may relate to your present assignment. Answer the question but do not go into an extended discussion. You are being examined for a *new* job, not your present one. As a matter of fact, try to phrase *all* your answers in terms of the job for which you are being examined.

G. BASIS OF RATING

Probably you will forget most of these "do's" and "don'ts" when you walk into the oral interview room. Even remembering them all will not insure you a passing grade. Perhaps you did not have the qualifications in the first place. But remembering them *will* help you to put your best foot forward, without treading on the toes of the board members.

Rumor and popular opinion to the contrary notwithstanding, an oral board wants you to make the best appearance possible.They know you are under pressure -- but they also want to see how you respond to it as a guide to what your reaction would be under the pressures of the job you seek. They will be influenced by the degree of poise you display,the personal traits you show,and the manner in which you respond.

EXAMINATION SECTION

EXAMINATION SECTION

TEST 1

DIRECTIONS: Each question or incomplete statement is followed by several suggested answers or completions. Select the one that *BEST* answers the question or completes the statement. *PRINT THE LETTER OF THE CORRECT ANSWER IN THE SPACE AT THE RIGHT.*

1. Backlash in gears which mesh together 1.____
   A. tends to reduce noise
   B. decreases with wear
   C. is the "play" when gears are reversed
   D. reduces the need for lubrication

2. A tapered shaft six inches long has end diameters of 2 inches and 1 1/2 inches. The rod has a taper per foot of 2.____
   A. 1 1/2 inches
   B. 1 inch
   C. 3/4 of an inch
   D. 1/2 of an inch

3. If a 1 1/2 inch circular shaft is required to be machined to a tolerance of plus or minus three thousandths of an inch, a circular shaft will have to be *SCRAPPED* if it has a diameter of 3.____
   A. 1.496 inches
   B. 1.497 inches
   C. 1.502 inches
   D. 1.506 inches

4. The area of a circular plate will be reduced by 5% if a sector removed from it has an angle of 4.____
   A. 18 degrees
   B. 24 degrees
   C. 32 degrees
   D. 60 degrees

5. If a 4 1/16 inch shaft wears six thousandths of an inch, the new diameter will be 5.____
   A. 4.0031 inches
   B. 4.0565 inches
   C. 4.0578 inches
   D. 4.0605 inches

6. A pinion with 12-teeth engages a gear rack having 6-teeth per inch. When the rack has moved two inches, the pinion will have made 6.____
   A. 1/4 of a revolution
   B. 1/2 of a revolution
   C. 1 revolution
   D. 2 revolutions

7. The gases *MOST* commonly used together in the torch cutting of steel are oxygen and 7.____
   A. nitrogen B. hydrogen C. acetylene D. methane

8. The *MOST* important precaution which should be taken if a number of men are in a small enclosure constantly using carbon tetrachloride for cleaning parts is to 8.____
   A. have plenty of fire extinguishers around because of the inflammability of the liquid
   B. make certain the enclosure has adequate ventilation
   C. have the men wear rubber gloves because this liquid is corrosive
   D. make certain the men wear safety goggles

9. A material used for lining bearings in order to reduce friction is 9.____
   A. babbitt
   B. cast iron
   C. carbon steel
   D. malleable iron

10. One of your men tells you that a small electric hand drill he has been using has burned out. On inspecting the drill you notice the man was using a 3/4" drill bit having a 1/4" shank to fit the drill. In this case you should tell him 10.____
   A. that he should have used a 3/4" drill with a 3/4" shank in this drill
   B. that the failure was probably due to pressing too lightly on the drill while drilling
   C. that failure under these conditions is sometimes due to overloading the drill by excessive pressure while drilling
   D. nothing, but allow him to proceed with the job using a similar drill in good condition

11. The designation 18-4-1 as it refers to steel is indicative of a type of 11.____
   A. low carbon steel
   B. corrosion resistant steel
   C. high speed tool steel
   D. forging quality steel

12. The sum of 1 1/2, 2 1/32, 4 3/16 and 1 7/8 is, *MOST NEARLY*, 12.____
   A. 9.593
   B. 9.625
   C. 9.687
   D. 10.593

13. The cone angle of the countersinking tool for flat head machine screws is, *MOST NEARLY*, 13.____
   A. 60°
   B. 82°
   C. 90°
   D. 118°

14. It is *GENERALLY* preferred that, for cutting brittle materials such as gray cast iron or magnesium, the cone angle on a drill should be 14.____
   A. less than 85°
   B. between 90° and 118°
   C. between 118° and 135°
   D. between 135° and 180°

15. In order to hold a round workpiece securely to a work bed of a milling machine, it is *PREFERABLE* to mount the workpiece 15.____
   A. on an angle plate bolted to the table
   B. in a three jaw chuck bolted to the table
   C. in V-blocks clamped to the table
   D. on a rotary attachment bolted to the table

16. The designation 51-1/2 A 18-2 on a milling machine arbor indicates an arbor of 16.____
   A. #50 taper, 1 1/2 inches diameter, style A, 18 inches length, and a No. 2 bearing
   B. #51 1/2 taper, style A, 18 1/8 inches length
   C. #51 1/2 taper, style A, 18 1/4 inches length with a pilot support
   D. #18 taper, 2 inches diameter, style A, 51 1/2 inches long

17. A plate 10 inches long is to have four 3/4 inch diameter holes drilled on 2 1/4 inch centers. In connection with this plate answer the following problem: 17.____

The two end holes are to have equal distances from the ends of the plate. The distance from the end of the plate to the center of the first hole should be

A. 1 1/2 inches B. 1 5/8 inches
C. 1 3/4 inches D. 2 inches

18. A typical application of a "follower rest" on a lathe is to support 18.____
A. a short workpiece for internal threading
B. the pilot bearing of a style "C" arbor
C. the taper attachment with the small end toward the tailstock
D. a long workpiece and prevent it from springing away from the tool

19. Of the following, the *MAJOR* advantage of an expansion reamer is that 19.____
A. holes can be progressively enlarged
B. one tool can be used for either hand or machine reaming
C. it can be easily adjusted to act as either a rose chucking or finishing reamer
D. it can be quickly adjusted to an exact size

20. "Loading and galling" in a drilling operation is *USUALLY* caused by using 20.____
A. drills with cutting lips with unequal angles
B. drills with improper web thinning
C. drills with cutting lips of unequal length
D. wrong type of drill or inadequate cutting fluid application

21. To determine hardness of a material over a very minute area, such as a microhardness test, one would employ the 21.____
A. Moh's hardness scale B. Vichers hardness test
C. Tukon tester D. Durometer

22. The device used to determine the hardness of a metal by measuring the resistance it offers to the penetration of a steel ball under pressure is: 22.____
A. Brinell B. Rockwell
C. Scleroscope D. Brush surface analyzer

23. A heat treatment applied to steel that consists of heating the steel slightly above its critical range and cooling very slowly under controlled conditions, is called 23.____
A. normalizing B. case hardening
C. annealing D. tempering

24. A fixed gage for measuring external dimensions having one gaging surface fixed while the opposite surface may be adjusted over a small range and locked in any desired position, is called 24.____
A. snap gages B. ring gages
C. air gages D. comparators

25. The height of a surface gage is usually set by using a/an 25.___
A. sine bar B. angle plate
C. protractor D. combination square

26. The one of the following that is *NOT* generally used to lubricate cutting tools is: 26.___
A. mineral oil B. soda water
C. kerosene D. lard oil

27. The standard marking system 32A50-J5VBE on a grinding wheel indicates, *MOST NEARLY* that the wheel 27.___
A. has a medium grain size B. is silicate bonded
C. has a fine grain size D. abrasive is corundum

28. When dressing a grinding wheel the *MOST* commonly used type of wheel dresser is the 28.___
A. mechanical dresser B. abrasive wheel
C. abrasive stick D. silver point

29. Allowance fit between plain cylindrical parts is *BEST* described by which one of the following: 29.___
A. Total permissible variation of size
B. Applicable maximum and minimum size
C. Limits of size prescribed for clearance only
D. Minimum clearance or maximum interference between mating parts

30. The one of the following that is *USUALLY* responsible for damaging milling cutters is: 30.___
A. high speeds
B. coarse feeds
C. dull cutters
D. poor rigidity of the work piece

31. The one of the following features that distinguishes a universal milling machine from a plain milling machine is: 31.___
A. cutting tool B. power feeds
C. less selection of feeds D. swivel table

32. For the rapid index plate of a Brown and Sharpe index head on milling machines, the number of equally spaced holes is 32.___
A. 16 B. 20 C. 24 D. 30

33. The diametral pitch of a spur gear with 40 teeth and a pitch diameter of 4.0 inches is 33.___
A. 5 B. 10 C. 20 D. 40

34. In a centerless grinder, the axial movement of the work piece for through-feed grinding is accomplished by the 34.___
A. end feed
B. work rest guide
C. regulating wheel
D. taper of the grinding wheel

35. In the honing of an internal cylinder, the *USUAL* method of tool control is to 35.____
A. combine a rapid rotation with an oscillatory motion
B. gradually force the stones against the wall by hydraulic pressure
C. move the head radially while rotating the tool slowly
D. combine a slow rotation with an oscillatory axial motion

36. A 2" in diameter shaft is making 67 RPM in a lathe. The cutting speed is, *MOST NEARLY*, 36.____
A. 28 fpm B. 30 fpm C. 35 fpm D. 40 fpm

37. On an engine lathe, the cutting edge of a boring tool should be set 37.____
A. 1/32" above center
B. on center
C. 1/16 below center
D. 1/16" above center

38. The Morse Standard taper is, *APPROXIMATELY*, 38.____
A. .250 inch per foot
B. .500 inch per foot
C. .625 inch per foot
D. .750 inch per foot

39. To cut a right hand external American National Standard thread form on a lathe, the compound rest should be swiveled to an angle of 39.____
A. 70 degrees
B. 60 degrees
C. 32 degrees
D. 29 degrees

40. Which of the following tools would be used to enlarge a 1 1/2" deep, 3/4" diameter hole to a 7/8" concentric diameter section for the upper 1/2" of its depth: 40.____
A. expansion reamer
B. counterbore
C. countersink
D. broach

KEY (CORRECT ANSWERS)

1.	C	11.	C	21.	C	31.	D
2.	B	12.	A	22.	A	32.	C
3.	A	13.	B	23.	C	33.	B
4.	A	14.	B	24.	A	34.	C
5.	B	15.	C	25.	D	35.	D
6.	C	16.	A	26.	A	36.	C
7.	C	17.	B	27.	A	37.	B
8.	B	18.	D	28.	A	38.	C
9.	A	19.	D	29.	D	39.	D
10.	C	20.	D	30.	A	40.	B

EXAMINATION SECTION

TEST 1

DIRECTIONS: Each question or incomplete statement is followed by several suggested answers or completions. Select the one that BEST answers the question or completes the statement.

1. Screw stock is USUALLY an alloy of
   A. copper, zinc, and nickel
   B. copper, zinc, and tin
   C. copper, zinc, and lead
   D. copper and zinc

2. The MOST desirable alloy steel for use in automobile springs is
   A. chromium-molybdenum
   B. silicon-manganese
   C. chromium-nickel
   D. manganese-molybdenum

3. An alloy of approximately 2/3 nickel and 1/3 copper (with small amounts of iron, manganese, and silicon) is COMMONLY referred to as
   A. Muntz metal
   B. Inconel
   C. Nickel silver
   D. Monel

4. When using the "thread dial" in the cutting of a thread on a lathe it is NOT proper to close the split nut at
   A. any un-numbered line on the dial for a 5 1/4 per inch thread
   B. any odd-numbered line on the dial for a 10 1/2 per inch thread
   C. any line on the dial for a 6 per inch thread
   D. any numbered line on the dial for a 7 per inch thread

5. On a modern lathe the "constant surface speed control" separately controls the speed of the
   A. cross feed
   B. lead screw
   C. spindle
   D. feed rod

6. In a finish grinding operation on cast iron, the surface speed of the work should be MOST NEARLY
   A. 60 feet/min.
   B. 110 feet/min.
   C. 150 feet/min.
   D. 200 feet/min.

7. The size of a standard hardened and ground lathe mandrel is USUALLY marked
   A. on the large end
   B. on both ends
   C. on the small end
   D. near the middle

8. A half-center used in facing work looks MOST NEARLY like

A. B. C. D.

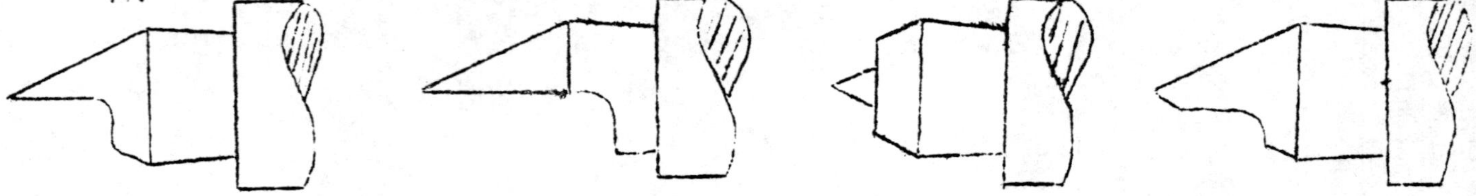

9. Assume that a cross feed screw on a special lathe has 10 threads per inch and the scale on the cross feed screw is divided up into 50 equal divisions. If it is desired to take a second cut on a stock which will reduce the stock diameter by .012", the feed should be moved in MOST NEARLY
   A. 12 divisions
   B. 6 divisions
   C. 3 divisions
   D. 1 1/2 divisions

10. For a very fine polishing on a lathe, it is MOST desirable to use a
   A. high spindle speed, oiled crocus cloth, and a heavy pressure on the work
   B. medium spindle speed, oiled crocus cloth, and a light pressure on the work
   C. low spindle speed, fine oiled emery cloth, and a light pressure on the work
   D. low spindle speed, fine dry emery cloth, and a very light pressure on the work

11. On a lathe, when extreme accuracy for centering a moderately heavy piece of work is needed, it is MOST desirable to use a
   A. drill chuck
   B. two-jaw chuck
   C. three-jaw self-centering universal chuck
   D. four-jaw independent chuck

12. Of the following, the piece of equipment NOT generally used in lathe work, is a(n)
   A. angle plate
   B. boring bar
   C. swage block
   D. gang mandrel

13. The MOST NEARLY CORRECT of the following statements about "lapping" is that it is a process in which a surface is abraded by
   A. rubbing dry with a harder material
   B. rubbing dry with a softer material
   C. rubbing with a softer material charged with fine abrasive particles in a vehicle such as grease
   D. rubbing with a harder material charged with fine abrasive particles in a vehicle such as oil

14. In connection with centerless external grinding, the statement which is MOST NEARLY CORRECT is that
    A. the grinding wheels are approximately of the same size
    B. the grinding wheels revolve at the same speed
    C. the shafts of the grinding wheels are not parallel
    D. the grinding wheels run in opposite directions

15. Assume that a grinding wheel is marked "C14-N12-S23" by the manufacturer. This marking MOST LIKELY indicates
    A. the manfacturer's inventory control number
    B. a vitrified bond silicon carbide wheel with fine grain size
    C. a silicate bonded aluminum oxide wheel with coarse grain size
    D. a silicate bonded silicon carbide wheel with coarse grain size

16. If, in the three-wire-method of measuring pitch diameters of screw threads, the actual micrometer reading over the wires is greater than the properly calculated reading, it is likely that the
    A. wires used are oversized
    B. thread needs further cutting
    C. thread is satisfactory
    D. thread is cut too deep

17. To produce a true flat on an iron surface it is BEST to use a
    A. planer B. shaper C. scraper D. milling machine

18. In lay-out work, the LEAST useful means for making scribed lines on steel more distinct is the use of
    A. bluing B. copper sulphate
    C. dilute acetic acid D. colored lacquer

19. The GREAT advantage of electron beam machining is that
    A. it can be used on massive work since it removes material slowly
    B. it can be used on miniaturized work since it removes material slowly
    C. the equipment needed costs much less than the equipment needed for electrical discharge machining
    D. a steady beam can be used to drill holes, up to 1/2 inch in diameter, instantaneously in metal thicknesses up to 12 inches

20. One disadvantage of electrical discharge machining is that this method CANNOT be used when
    A. it is desired to remove material at a slow rate
    B. the material to be machined is an electrical conductor
    C. the material to be machined has a Brinell of over 400
    D. a surface finish of less than 500 microinches is required

21. When a thread is to be used to transmit a maximum amount of power with minimum friction it is MOST desirable to cut a(n)
    A. American National thread B. American National Acme threa
    C. 29 degree worm thread D. square thread

22. A safety lathe dog differs from an ordinary lathe dog in that the former has
    A. two diametral tails for positive drive
    B. a tail of soft metal which shears under dangerous load
    C. a large headed setscrew for secure clamping
    D. a headless set screw that does not protrude

23. Location of the center at the end of a piece of round stock is MOST easily accomplished by using
    A. a surface gage
    B. spring calipers
    C. dividers
    D. a center head

24. A "combination set" has a
    A. scale, square head, universal bevel, and a radius gage
    B. scale, bevel protractor, center head, and square head
    C. level, bevel protractor, radius gage, and center head
    D. scale, universal bevel, radius gage, and square head

25. A telescoping gage is USUALLY used for making
    A. measurements of long lengths
    B. measurements of large external diameters
    C. measurements with a comparator
    D. internal measurements

26. The recommended number of teeth per inch on a hacksaw blade that is to be used for cutting thin sheet metal and thin tubing is, MOST NEARLY,
    A. 32
    B. 24
    C. 18
    D. 14

27. Among the common types of files that are used in machine shops, the file that has two "safe" edges is known as a
    A. pillar file
    B. mill file
    C. hand file
    D. flat file

28. Trammel points CANNOT be used
    A. as inside calipers
    B. for scribing arcs from holes
    C. as a surface gage
    D. for scribing arcs from punch marks

29. A wiggler indicator is used PRIMARILY for
    A. measuring frequency of vibration
    B. indicating gear backlash
    C. centering work in machine tools
    D. determinations of surface finish

30. Toolmakers' buttons are USUALLY used for
    A. plugging inaccurately drilled holes
    B. Laying out holes accurately
    C. gaging hole sizes
    D. checking shallow hole depths

31. Among the following gages, the one which is used to check a specific profile of an object is called a
    A. "go" ring gage
    B. radius gage
    C. drill and wire gage
    D. "not go" plug gage

32. For average work, the proper cutting speed to use in a small power hacksaw should be, MOST NEARLY,
   A. 10 to 20 strokes per minute
   B. 3o to 40 strokes per minute
   C. 50 to 60 strokes per minute
   D. 80 to 100 strokes per minute

33. A sine bar is a tool that may be used for accurate
   A. boring
   B. centering of round stock
   C. torque measurement
   D. angle measurement

Question 34 is based on the sketch below.

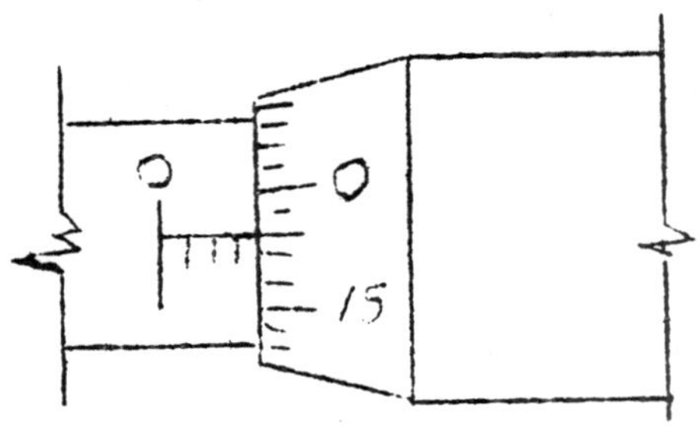

34. Assume that there were 50 threads per inch on a micrometer spindle (instead of the usual 40), and the thimble was graduated in 20 divisions (instead of the usual 25). With 50 divisions on the barrel (instead of the usual 40), the setting shown above, on such a micrometer, would indicate a reading of
   A. .078"
   B. .093"
   C. .096"
   D. .098"

Questions 35 is based on the sketch below showing part of a 3"-4" inside micrometer.

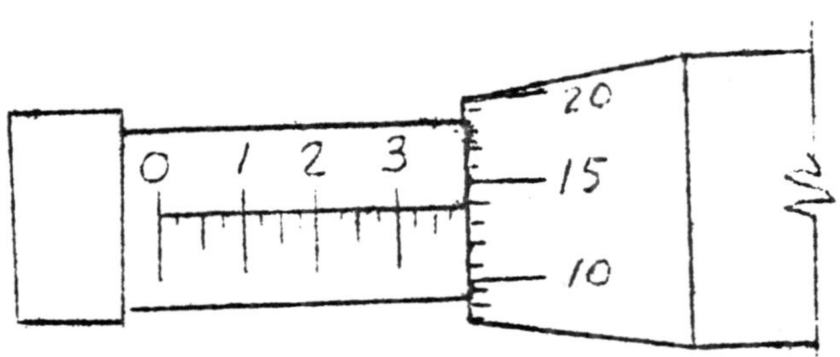

35. With a setting as shown above, the dimension being measured, is, MOST NEARLY,
   A. 3.989"
   B. 3.890"
   C. 3.611"
   D. 3.389"

36. A micrometer graduated to read to .0001" USUALLY has
   A. no vernier scale
   B. a vernier scale on the anvil
   C. a vernier scale on the barrel
   D. a vernier scale on the thimble

37. A profilometer is an instrument which may be used for
    A. tracing outlines on shop drawings
    B. measuring surface roughness
    C. controlling movement of the shaper toolslide
    D. measuring the areas of irregular figures

38. Assume that a tapered shank having a 0.600 inch per foot taper goes into a taper gage too far by 0.200 inch. From this it can CORRECTLY be said that the diameter of the taper is too small by
    A. 0.005" B. 0.010" C. 0.015" D. 0.020"

39. With reference to the measurement of hardness, a "Brale" is a
    A. hardened steel ball 1 centimeter in diameter
    B. Moh's scale of hardness of 3
    C. hardened steel ball 1 1/16 inch in diameter
    D. diamond cone

40. A Brinell number of 400 corresponds MOST NEARLY to a
    A. Rockwell "C" of 42 B. Rockwell "B" of 84
    C. Sclerescope number of 21 D. tensile strength of 84,000 psi

41. With reference to milling, the statement which is MOST NEARLY CORRECT is:
    A. Heavy down milling cuts can be made safely on any milling machine.
    B. Down milling results in poorest metal finishes.
    C. Down milling does not require as firm a clamping of work as in up milling.
    D. With hard surfaced materials, down milling gives greater productivity per sharpening.

42. Assume that on a milling machine fitted with a dividing head, 24 evenly spaced cuts are to be taken around the circumference of a work piece. With a 40 to 1 gear ratio and a 36 hole index circle, the number of holes to be indexed with plain indexing after each cut, is, MOST NEARLY,
    A. 20 B. 40 C. 60 D. 80

43. To be able to take two cuts on a piece of work in a milling machine so that the cuts are 30 minutes of arc apart, the minimum number of holes for plain indexing on an index circle fitted with a 40 to 1 gear ratio, should be, MOST NEARLY ,
    A. 8 B. 12 C. 18 D. 36

44. The designation "right=hand cutter" is the proper designation to use when referring to an
    A. end-mill cutter that rotates counter-clockwise when viewed from the front end of the spindle
    B. end-mill cutter that rotates clockwise when viewed from the front end of the spindle
    C. arbor-type cutter that rotates clockwise when viewed from the front end of the arbor
    D. arbor-type cutter that rotates counter-clockwise when viewed from the front end of the arbor

45. Assume that a jobbing shop is to submit a price for a contract involving 300 pieces of work. Assume that material costs 25 cents per piece, labor costs $3.75 an hour, and a lathe operator can complete 5 pieces in an hour. If overhead is 40% of material and labor costs and the profit is 10% of all costs, the submitted price for the entire job will be, MOST NEARLY,
   A. $315.12   B. $438.90   C. $450.00   D. $462.00

46. The following formula is used in connection with the three-wire method of measuring pitch diameters of screw threads:

$$G = \frac{0.57735}{N}$$

where G = wire size
N = number of threads per inch.

According to this formula, the proper size of wire for a 1"-8NC thread is, MOST NEARLY,
   A. .0722"   B. .7217"   C. .0072"   D. .0074"

47. A millimeter is 1/25.4 of an inch and there are 10 millimeters to a centimeter. If a piece of stock measures 127 centimeters long, the length of the stock in feet and inches would be, MOST NEARLY,
   A. 2 ft. 1 inch   B. 4 ft. 2 inches
   C. 8 ft. 4 inches   D. 41 ft. 8 inches

48. The MAIN purpose of the tang on a taper shank drill is to
   A. supply the negative action required to drive the drill
   B. serve as a holder when the drill is being sharpened
   C. provide the means for removing the drill by use of a drift
   D. relieve the stress on the drill in heavy use

49. The statement concerning drills which is, MOST NEARLY CORRECT, is:
   A. Taper shanks are available only on drills where the diameter is 1/2" or larger.
   B. Straight shank drills are available in sizes up to 3" diameter.
   C. Taper shanks are available for drills of all sizes.
   D. Straight shank drills are available in sizes up to 1/2" diameter.

Question 50 is based on the sketch below.

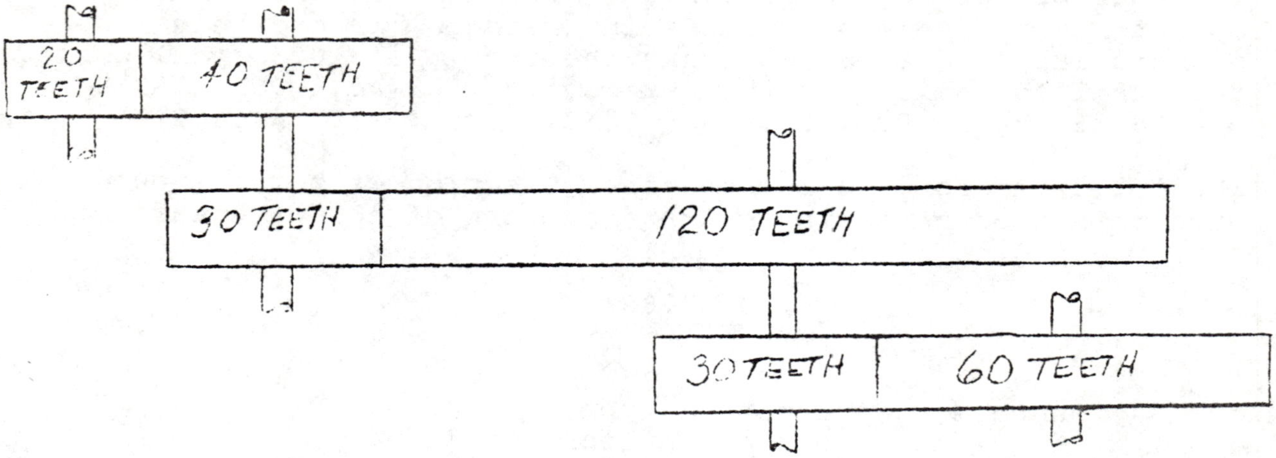

50. Assume that the number of teeth on each gear of a compound gear train is shown above. If the 60 tooth gear makes 60 clockwise revolutions per minute, the 20 tooth gear will MOST LIKELY make
A. 480 rpm B. 720 rpm C. 960 rpm D. 1200 rpm

KEY (CORRECT ANSWERS)

1.	C	11.	D	21.	D	31.	B	41.	C
2.	B	12.	C	22.	D	32.	C	42.	C
3.	D	13.	C	23.	D	33.	D	43.	C
4.	A	14.	C	24.	B	34.	A	44.	A
5.	C	15.	D	25.	D	35.	D	45.	D
6.	A	16.	B	26.	A	36.	C	46.	A
7.	A	17.	C	27.	A	37.	B	47.	B
8.	D	18.	C	28.	C	38.	B	48.	C
9.	C	19.	B	29.	C	39.	D	49.	D
10.	A	20.	D	30.	B	40.	A	50.	C

TEST 2

DIRECTIONS: Each question or incomplete statement is followed by several suggested answers or completions. Select the one that BEST answers the question or completes the statement.

1. With reference to twist drills, the statement MOST NEARLY CORRECT is :
   A. When a high speed drill is being ground dry, it should be dipped in water to cool it.
   B. To cut properly, one cutting lip should be ground longer than the other.
   C. The lip clearance should be approximately 12 degrees.
   D. The angle that the lips form with the centerline of the drill should be 69 degrees.

2. When drilling a hole in 1020 steel with a 3/4" HSS drill, the MOST desirable drill speed and feed would be MOST NEARLY
   A. 600 rpm and .003 ipr B. 900 rpm and .010 ipr
   C. 1200 rpm and .020 ipr D. 1500 rpm and .025 ipr

3. Assume that a lathe has a spindle speed of 144 revolutions per minute while 12 threads per inch are to be cut along the entire 6" length of a 2" diameter piece of stock. The MINIMUM time required to complete the first cut is,MOST NEARLY,
   A. 1 minute B. 45 seconds C. 30 seconds D. 15 seconds

4. A cam to give a lift of 2" is to be made by drilling an off-center hole in a 3" diameter circular disk so that the disk can be mounted on a shaft. The hole should be located
   A. 1/2" off center B. 1" off center
   C. on the circumference D. the job is impossible

5. The hole diameter for a shaft in a journal bearing has a dimension of $3.408"^{+.002}_{-.003}$ and the shaft dimension is $3.404"^{+.000}_{-.001}$.
   The limits of the clearance for the shaft in the bearing are, MOST NEARLY,
   A. .009" max. .001" min. B. .008" max. .001" min.
   C. .007" max. .002" min. D. .007" max. .001" min.

Question 6 refers to the sketch below which indicates a work piece to be manufactured as per indicated dimensions and tolerances.

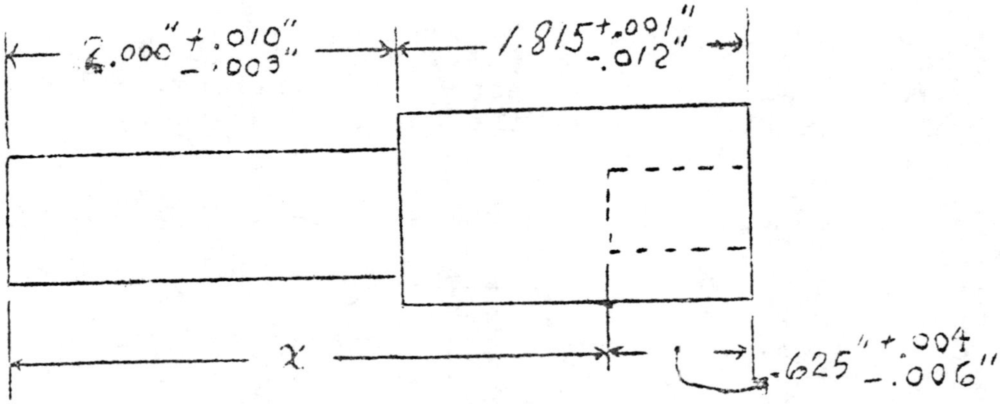

6. According to the above sketch, the MAXIMUM dimension for x is, MOST NEARLY,
   A. 3.250"   B. 3.254"   C. 3.267"   D. 3.276"

Question 7 is based on the sketch below. Assume that it is desired to drill four holes in a plate as shown in the diagram below.

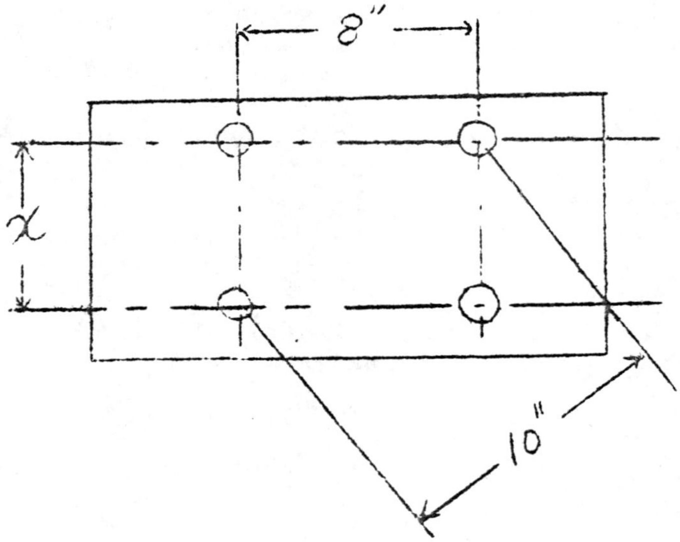

7. The dimension x is. MOST NEARLY,
   A. 5.62"   B. 5.81"   C. 6.00"   D. 6.25"

8. Assume that the following Webber blocks are made available to you:
.1003, .1004, .1007, .1009, .100, .105, .107, .112, .115, .128 , .130, .146, .250, .450,1.000.

Of the above blocks, the SMALLEST number of blocks you could use to establish the dimension of 1.3427" is
A. 4 blocks B. 5 blocks C. 6 blocks D. 8 blocks

Question 9 refers to the sketch below.

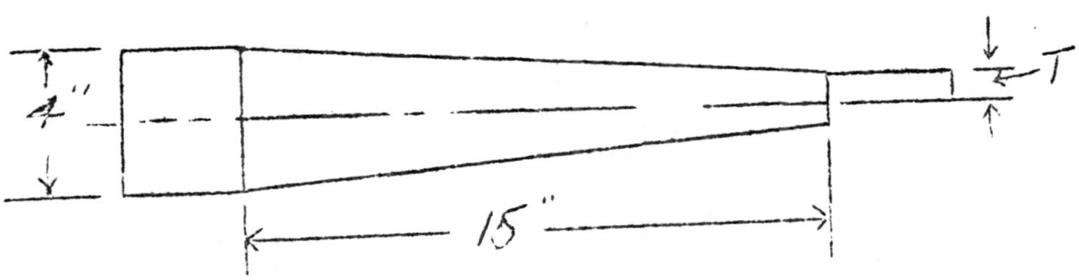

9. Assuming that the taper in the above drawing is one inch per foot, the dimension T will be, MOST NEARLY,
A. .750" B. 1.250" C. 1.375" D. 1.438"

Question 10 is based on the sketch below. Assume that the object pictured in the following three views is solid and made of steel.

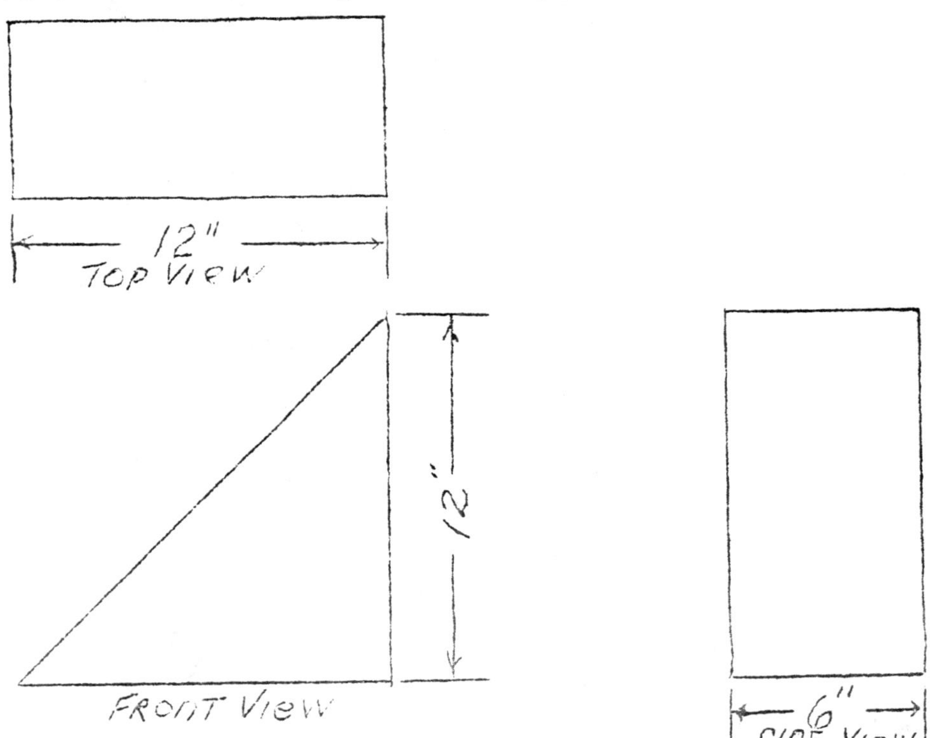

10. If steel weighs 480 lbs. per cu.ft., the weight of the above object will be, MOST NEARLY,
A. 70 lbs. B. 120 lbs. C. 146 lbs. D. 224 lbs.

11. In comparing mechanically and hydraulically driven shapers, the statement MOST NEARLY CORRECT is :
   A. Qucker return strokes can be obtained by mechanical drives.
   B. Hydraulic drives are less complicated and cheaper than mechanical drives.
   C. With hydraulic drives, cutting stroke velocities are more uniform than with mechanical drives.
   D. A shaper with hydraulic drive is more easily over-loaded.

12. In using a shaper to machine a 2" X 6" X 1" rectangular block, the PREFERRED order of machining the surfaces, where the 2" X 6" surfaces are the top and bottom and the 1" X 6" surfaces are the sides, is
   A. 1-side 1, 2-top, 3-bottom, 4-side 2
   B. 1-top, 2-bottom, 3-side 1, 4-side 2
   C. 1-top, 2-side 1, 3-side 2, 4-bottom
   D. 1-side 1, 2-side 2, 3-top, 4-bottom

13. With reference to formed tooth milling cutters, it can CORRECTLY be said that
   A. the cutters do not have any relief
   B. such cutters can be used only for milling simple contours
   C. when the faces of the teeth are sharpened, the contour of the cutting edge is unchanged
   D. they cannot be used in gang milling

14. Assume that specifications call for tightening a nut with a torque of 900 inch pounds. To accomplish this, the tangential force applied to the end of an 18 inch wrench would have to be MOST NEARLY
   A. 50 lbs.   B. 75 lbs.   C. 112.5 lbs.   D. 150 lbs

15. It is desired to connect two pipe fittings as shown below:

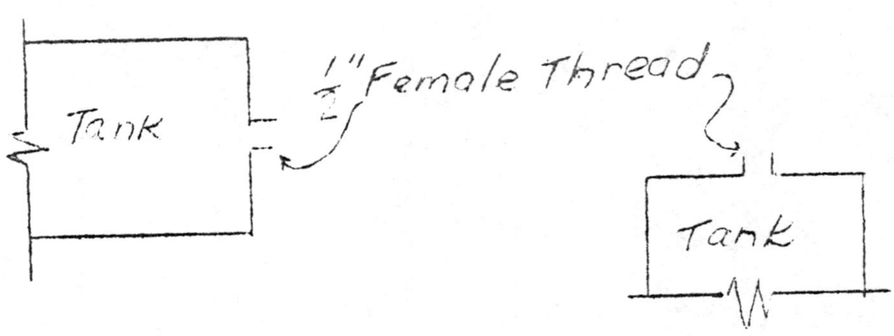

This could be accomplished by using a length of pipe threaded on both sides in addition to a
   A. coupling and a straight cross
   B. nipple, a 90° elbow and a union
   C. nipple, a 90° street elbow and a union
   D. nipple, a coupling and a union

16. In requisitioning sheets of very coarse abrasive cloth, the requested grit size should be MOST NEARLY
    A. 36 B. 80 C. 180 D. 320

17. Fullers, flatters, and swages are tools USUALLY used in
    A. casting (foundry work) B. forging
    C. plumbing D. welding

18. With reference to files, the statement MOST NEARLY CORRECT is:

    A. A new file should be broken in by using it first on brass or bronze.
    B. Files should be oiled when filing cast iron.
    C. The body and tang of a file are both very brittle.
    D. Rubbing chalk between the teeth of a file will cause pinning.

19. An N-16 thread, as designated in the American National Thread System, indicates that the thread has a
    A. 1/16" crest B. 16" lead
    C. half angle of 16° D. 1/16" pitch

20. The electrolytic process in which the thickness of the protective layer of aluminum oxide on aluminum is increased is called
    A. metalizing B. electroplating
    C. anodizing D. parkerizing

21. Ferrous castings are USUALLY repaired by the process known as
    A. brass soldering B. dip brazing
    C. braze welding D. resistance brazing

22. Soldering with a soldering iron requires the use of a flux on the parts to be joined. The material used LEAST frequently as a flux is
    A. sal ammoniac B. zinc chloride
    C. rosin D. none of these

23. The LEAST desirable method of removing a broken tap from a hole is to try to
    A. use a tap "extractor"
    B. jar the tap loose with a punch
    C. break the tap into small pieces
    D. drill the tap out

24. In order to relieve stresses and to induce softness in steels, the MOST effective heat treatment that can be used is called
    A. annealing B. carburizing C. quenching
    D. nitriding

25. When quenching is resorted to in the heat treatment of steel, it is UNDESIRABLE to use
    A. volatile oils B. sodium hydroxide
    C. potassium hydroxide D. brine

26. The compound MOST frequently used in casehardening is
    A. potassium hydroxide
    B. potassium cyanide
    C. sodium sulphate
    D. cryolite

27. The terms "Standard", "ratchet", "offset", "clutch head", USUALLY refer to
    A. box wrenches
    B. screwdrivers
    C. pipe wrenches
    D. spanners

28. For the purpose of filing a slot or keyway, it would be DESIRABI to use a
    A. pillar file
    B. flat file
    C. mill file
    D. taper file

Questions 29-30.

Questions numbered 29 to 30 are to be answered in accordance with the paragraph below.

"When one is making the selection of grinding wheel specifications, the first variable factor to consider is the wheel speed, which influences the grade and the bond of the wheel. It is recommended that the grade should be determined in this way: the higher the wheel speed with relation to work speed, the softer the wheel should be. When, for any reason, the wheel speed is reduced, then it may be expected that the wheel will wear faster, but this can be overcome by choosing a wheel of a harder grade, assuming, that the grade was correct for the initial speed."

29. It can be said that the MOST important piece of information in the above paragraph is
    A. The higher the relative wheel speed, the softer should be the wheel.
    B. Wheel speed is a variable factor.
    C. At low speeds wheels wear rapidly.
    D. When a wheel slows down, it should be replaced by a harder grade.

30. According to the above paragraph, NO indication is made that
    A. there are other factors to be considered besides speed
    B. hard wheels at low speed wear faster than soft wheels at high speed
    C. the lower the speed the harder should be the grade
    D. the selection of the bond of the wheel is affected by speed

31. The carriage or saddle on an engine lathe is BEST described as the
    A. movable part which slides over the ways between the headstock and tailstock
    B. movable casting located opposite the headstock on the ways
    C. object that directly supports the cutting tool
    D. two heavy metal sides located lengthwise with ways or V's formed upon them

32. In setting up a milling machine for "rapid or direct" method of indexing it is necessary to
    A. connect the index plate to the spindle by means of a gear train
    B. disengage the worm and gear
    C. index by means of the index plate and index crank
    D. change the size of gear on the worm

33. The gears BEST suited to transmit motion between shafts whose center lines intersect are
    A. spur gears   B. bevel gears
    C. helical gears   D. spiral gears

34. Coolants made of soluble oils or emulsions should NOT be used for machining
    A. bronze   B, aluminum   C. magnesium   D. alloy steels

35. When using high-speed tool bits, ordinary machine steel is generally turned on a lathe at a speed, in feet per minute, of MOST NEARLY
    A. 30 to 45   B. 50 to 75   C. 90 to 150   D. 175 to 225

36. The speed of the cutting stroke, for planing steel castings, roughing, and finishing, in feet per minute, is MOST NEARLY
    A. 30 to 40   B. 50 to 75   C. 100 to 150   D. 160 to 200

37. Planing speeds are lower than cutting speeds for turning, MAINLY because in planing
    A. higher speeds would require special coolants
    B. it is not practicable to use an ample supply of cutting fluid
    C. special tool bits would be required for the higher speeds than those for turning
    D. the table moves only in one direction

38. A 1/2" diameter screw, 13 threads per inch, American Coarse Thread Series, right hand, and loose fit is USUALLY specified on a drawing as
    A. 1/2" - 13NC - 4 - RH   B. 1/2" - 13NC - 3 - LH
    C. 1/2" - 13NC - 2   D. 1/2" - 13NC - 1

39. An arbor is a device for
    A. holding parts to be machined
    B. holding and driving cutting tools
    C. holding parts on a magnetic grinder
    D. enlarging small holes in thin metals

40. Five graduations on the barrel of a micrometer caliper indicates an opening of MOST NEARLY
    A. .005"   B. .025"   C. .125"   D. .135"

41. Seventeen graduations on the barrel of a micrometer caliper and nineteen graduations on the thimble indicates an opening of MOST NEARLY
    A. .545"   B. .444"   C. .394"   D. .339"

42. Increasing the speed of the shaper after completing the roughing cuts results in
    A. chattering
    B. a better finish
    C. increasing internal stresses
    D. warpage

43. A device used for supporting irregular shaped objects in a steady rest of a lathe is called a
    A. hold down
    B. cube or box parellel
    C. parallel
    D. cat head

44. In producing a large number of small screws, studs or bolts on a lathe, the PROPER chuck to use for this type of work is the
    A. independent chuck
    B. universal chuck
    C. draw-in chuck
    D. combination chuck

45. The proper side rake angle of a cutting tool for turning soft brass is MOST NEARLY
    A. 6° B. 20° C. 30° D. 35°

46. In the statement, "The Machinist is machining a bearing housing of conventional design", the word conventional means MOST NEARLY
    A. complicated B. superior C. new D. common

47. In the statement, "When turning a piece of tenacious metal on a lathe, a lubricant is used to prevent excessive friction by conducting the heat away", the word tenacious means MOST NEARLY
    A. annealed B. soft C. tough D. coarse

48. In the statement, "In a particular shop a Machinist is assigned to the task of coordinating various machining operations", the word coordinating means MOST NEARLY
    A. repairing B. replacing C. testing D. scheduling

49. In the statement, "The Machinist made an insignificant error", the word insignificant means MOST NEARLY
    A. serious B. accidental C. minor D. hidden

50. "Some machinists have the faculty of knowing when there is work to be done and do not have to be stimulated into doing it." These machinists may be said to possess
    A. neatness B. individuality C. obedience D. initiative

KEY (CORRECT ANSWERS)

1.	C	11.	C	21.	C	31.	A	41.	B
2.	A	12.	C	22.	A	32.	B	42.	B
3.	C	13.	C	23.	D	33.	B	43.	D
4.	B	14.	A	24.	A	34.	C	44.	C
5.	D	15.	C	25.	A	35.	C	45.	A
6.	C	16.	A	26.	B	36.	A	46.	D
7.	C	17.	B	27.	B	37.	B	47.	C
8.	A	18.	A	28.	A	38.	D	48.	D
9.	C	19.	D	29.	A	39.	B	49.	C
10.	B	20.	C	30.	B	40.	C	50.	D

EXAMINATION SECTION

TEST 1

DIRECTIONS: Each question or incomplete statement is followed by several suggested answers or completions. Select the one that BEST answers the question or completes the statement.

1. When using differential indexing on a milling machine, the index-plate is USUALLY rotated by gearing connected with the
   A. radial arm
   B. dividing-head spindle
   C. universal
   D. foot stock

2. A piece of round stock is 8 inches long. A taper of 3/4 inch per foot is required on the stock. The lathe tailstock center should be set over from the centerline position
   A. 3/16 inch B. 1/4 inch C. 5/16 inch D. 3/8 inch

3. In repairing a particular pump, it was necessary for a machinist to make four studs from a steel bar. Assuming the length of each stud measures 2 3/16", 1 11/32", 2 1/4" and and 1 7/8", and 1/8" waste is allowed for each cut taken and 1/32" is allowed for facing each end, then the length of rod needed is MOST NEARLY
   A. 8 7/16" B. 8 9/32" C. 7 31/32" D. 8 3/32"

4. Which one of the following statements concerning a lathe is MOST NEARLY CORRECT?
   A. The threading mechanism should be used for general "feeding".
   B. The live center is usually hard, and the dead center soft.
   C. The axis of the main spindle determines the centerline of the lathe.
   D. The tailstock slide has no short transverse adjustment.

5. The part of a lathe that contains the gears and clutches for transmitting motion from the feed rod to the carriage is called the A. Saddle B. Apron C. Headstock D. Cross slide

6. Regarding planers, the statement MOST NEARLY CORRECT is:
   A. Hardness of material is the most important factor in planing cast iron and ordinary steels.
   B. A deep cut with a light feed is better than a light cut and heavy feed.
   C. As depth of cut is increased, the work speed should be increased.
   D. By using carbide tools, it is advantageous to increase the feed.

7. When grinding crankpins, it is necessary to hold the main crankshaft in an offset position so that the axis of the crankpin to be ground will coincide with the axis of rotation or centerline of the grinding machine. In this particular case the amount of offset is equal to
   A. 1/4 the throw of the crank
   B. 1/2 the throw of the crank
   C. 3/4 the throw of the crank
   D. the full throw of the crank

Questions 8-9.

DIRECTIONS: Questions 8 and 9 are based on the following quotation:

"The wheels used for internal grinding should generally be softer than those used for other grinding operations, because the contact area between the wheel and work is comparatively large. A soft wheel that will cut with little pressure should be used to prevent springing the spindle. The grade of the wheel depends upon the character of the work and the stiffness of the machine, and, where a large variety of work is being ground, it may not be practicable to have an assortment of wheels adapted to all conditions. By adjusting the speed, however, a wheel not exactly suited to the work in hand can often be used. If the wheel wears too rapidly, it should be run faster, and, if it tends to glaze, the speed should be diminished."

8. On the basis of the above quotation only, it may BEST be said that
   A. the type and grade of wheel are independent of the sturdiness of the machine
   B. by increasing the wheel speed, parts can easily be internally ground
   C. wheels used for outside grinding usually have a smaller contact area between the wheel and work
   D. to carry on hand an assortment of wheels for all conceivable internal grinding jobs is economical

9. On the basis of the above quotation only, it may BEST be said that
   A. in general, if a wheel wears too rapidly, the speed should be decreased
   B. by decreasing the wheel speed, a wheel not quite appropriate for the job may sometimes be used
   C. where a large variety of work is being ground, the grade of wheel depends on the diameter of the wheel
   D. if a wheel tends to glaze, it should run faster

10. A three-inch diameter steel bar two feet long weighs MOST NEARLY (Assume steel weighs 480 lbs./cu.ft.)
   A. 48 lbs. B. 58 lbs. C. 68 lbs. D. 78 lbs.

11. Which one of the following statements concerning the numbering system used with Morse standard taper shanks is MOST NEARLY CORRECT?
    A. The taper per inch does not always increase with each larger taper number.
    B. The tapers are numbered from 1 to 7 inclusive.
    C. The taper per foot is 1/12 that of the taper per inch.
    D. Morse tapers are not generally used on the shanks of twist drills.

12. The micrometer equipped with a vernier scale USUALLY has an accuracy of
    A. .001" B. .0001" C. .00001" D. .00002"

13. A cracked malleable iron casting can BEST be repaired by welding or brazing it with rods made of
    A. cast iron B. steel C. bronze D. white metal

14. The flexible couplings of a pumping-unit (centrifugal pump and motor) are BEST aligned with the use of a
    A. straight edge and feeler gage
    B. dial indicator and feeler gage
    C. tachometer and feeler gage
    D. ram and feeler gage

15. To get the best performance from a steel twist drill, which is to be used on heat treated steels or drop forgings such as automobile connecting rods, the cutting edges of the drill should be ground so that the angle, in degrees, included between the two lips is MOST NEARLY
    A. 60 B. 90 C. 118 D. 125

16. Stay bolts are usually installed to prevent flexing of boiler heads. These stay bolts
    A. are welded to the heads
    B. have a fusible plug at one end
    C. are made of non-ferrous material
    D. are usually threaded rods

17. The cutting stroke speed of a shaper ram, in feet per minute, having a stroke of 8 inches and making 54 strokes per minute, is MOST NEARLY
    A. 40 B. 80 C. 50 D. 60

18. A device used to take up slack in cables and tie rods is called a A. Turnbuckle B. Clevis C. Collar screw D. Wing nut

19. A lathe with a lead number of 8, has a 24 tooth gear on the stud and a 48 tooth gear on the screw. This lathe is geared to cut
    A. 16 threads/in. B. 4 threads/in.
    C. 32 threads/in. D. 8 threads/in.

20. The diametral pitch of a gear usually indicates
    A. the size of the tooth
    B. the type of tooth
    C. the number of teeth per inch of circular pitch
    D. the number of teeth per inch of root circle

21. Draw filing can BEST be done to get a smooth finish with the use of a
    A. double cut file
    B. second cut file
    C. single cut file
    D. cross cut file

22. A drift is a tool used to
    A. drive taper pins and cotter pins
    B. align holes for bolts and rivets
    C. mark layout lines with small indentations
    D. drilling stone or concrete

23. A pawl is a device which is USUALLY found on a
    A. lathe
    B.. shaper
    C. drill press
    D. hand grinder

24. The American Standard Taper Pipe Thread is made with a taper of
    A. 1/4" per foot
    B. 3/8" per foot
    C. 1/2" per foot
    D. 3/4" per foot

25. The distance a screw thread advances axially in one turn is called
    A. pitch
    B. lead
    C. angle of thread
    D. crest

26. The chip pressure, in pounds, on a single-pointed tool, when machining cast iron using a feed of .012" per revolution and taking a cut 1/4" deep, would be MOST NEARLY (See data below.)
    A. 130
    B. 390
    C. 330
    D. 30

Assume that the chip pressure formula is

$$P = K \times f \times d \text{ with } K \text{ for cast iron} = 132{,}000$$
$$f = \text{feed/rev.}$$
$$d = \text{depth of cut.}$$

27. The size of a thread on a bolt could be determined by using a
    A. Snap gage
    B. Screw pitch gage
    C. "Go" gage
    D. Ring gage

28. A cold chisel is a very dangerous tool when
    A. the cutting edge is too sharp
    B. used with a ball peen hammer
    C. used with a bronze-faced hammer
    D. the head is mushroomed

29. A pair of bevel gears with the same number of teeth and with shafts at right angles to each other is called
    A. spiral bevel gears
    B. spur gears
    C. parallel gears
    D. miter gears

30. The MAIN object of "drawing" the temper in a piece of steel is to reduce its
    A. strength   B. malleability   C. ductility   D. brittleness

31. The tap drill for a 10/24 American National Coarse Thread is MOST NEARLY
    A. No. 36   B. No. 33   C. No. 26   D. No. 21

32. If a pulley rotates at 175 R.P.M. and is belt driven by a 12" diameter pulley which rotates at 275 R.P.M., then the diameter of the driven pulley is MOST NEARLY
    A. 15"   B. 19"   C. 16"   D. 23"

33. A snap gage is USUALLY used in connection with
    A. holes   B. threads   C. sine bars   D. outside diameters

34. In the formation of gear teeth, one of the systems in general use today is the "Involute System." The pressure angle for this particular system is USUALLY
    A. 12.5°   B. 13.5°   C. 20.0°   D. 14.5°

35. Which one of the following steels is USUALLY case hardened?
    A. high speed steels   B. cold rolled steels
    C. alloy steels   D. chromium steels

36. In negative rake cutting on a lathe, the back rake surface of the tool
    A. is held 90° to the movement of the work
    B. points in the direction of work movement
    C. points opposite to the work movement
    D. causes material to be sheared from the work

37. When turning rough forgings or steel castings on a lathe using a sintered carbide tool, the tool should be ground with a
    A. zero back rake   B. positive back rake
    C. negative back rake
    D. negative side rake of 15° to 20°

38. If a supervisor states that a particular type of breakdown on an automotive engine is general, he means it is MOST NEARLY
    A. timely   B. unimportant   C. obvious   D. prevalent

39. The type of machine used for milling irregular slots, contours, etc., on parts of machines or instruments, is called a
    A. Brinell   B. Rockwell
    C. Pantograph   D. Scleroscope

40. The part of a truck that allows one rear wheel to rotate faster than its mate when going around a corner is called the
    A. transmission   B. universal
    C. differential   D. crankshaft

41. A hacksaw blade with 14 teeth per inch is BEST used for sawing
    A. structural steel
    B. aluminum (#20 B.& S. gauge sheets)
    C. brass (#22 B.& S.gauge sheets)
    D. high speed steel

42. The lead of a screw that has a double thread is MOST NEARLY
    A. one-half the pitch
    B. one-quarter of the pitch
    C. twice the pitch
    D. four times the pitch

43. To finish the bottoms of narrow slots, the file BEST used is the
    A. taper B. square C. Vixen D. flat

44. The thread that has a 45° thread angle is COMMONLY known as the
    A. Acme B. Square C. Buttress
    D. Brown & Sharpe Worm

45. To get a smooth finish, draw filing can BEST be done by a machinist with the use of a
    A. double cut file B. second cut file
    C. single cut file D. cross cut file

46. The amount of material USUALLY left for finishing when filing on a lathe is MOST NEARLY
    A. .001" B. .004" C. .005" D. .007"

47. V-blocks and clamps are USUALLY used to hold
    A. square stock B. flat stock
    C. brass stock D. round stock

48. Of the following, when a fast cutting file is required, the able machinist should use a
    A. single cut bastard B. single cut smooth
    C. double cut bastard D. double cut smooth

49. The chisel used for chipping filleted corners and concave surfaces is a
    A. Round point B. Flat
    C. Diamond point D. Cape

50. The hand tap used to start a thread is called a
    A. plug tap B. taper tap
    C. bottoming tap D. drive tap

KEY (CORRECT ANSWERS)

1.	B	11.	A	21.	C	31.	C	41.	A
2.	B	12.	B	22.	B	32.	B	42.	C
3.	B	13.	C	23.	B	33.	D	43.	B
4.	C	14.	A	24.	D	34.	D	44.	C
5.	B	15.	D	25.	B	35.	B	45.	C
6.	B	16.	D	26.	B	36.	B	46.	A
7.	D	17.	A	27.	B	37.	C	47.	D
8.	C	18.	A	28.	D	38.	D	48.	C
9.	B	19.	A	29.	D	39.	C	49.	A
10.	A	20.	A	30.	D	40.	C	50.	B

TEST 2

DIRECTIONS: Each question or incomplete statement is followed by several suggested answers or completions. Select the one that BEST answers the question or completes the statement.

1. High spots on a surface plate which has been in use for some time are BEST removed by means of a
   A. router B. sander C. grinding wheel D. scraper

2. The CORRECT lip clearance angle to grind on a drill is GENERALLY
   A. 8° B. 12° C. 18° D. 22°

3. The APPROXIMATE drill speed, in feet per minute, for drilling cast iron is
   A. 50 B. 150 C. 200 D. 250

4. For drilling in steel, the angle that the lips or cutting edges should make with the axis of the drill is MOST NEARLY
   A. 39° B. 49° C. 59° D. 72°

5. The number of threads per inch for a screw with a pitch of .375 inches is
   A. 2 1/3 B. 2 1/4 C. 2 2/3 D. 2 1/2

6. The MAJOR thread diameter of a number 10 machine screw is
   A. .164" B. .175" C. .190" D. .215"

7. The tap drill size required for an 8-32 thread is number
   A. 40 B. 35 C. 21 D. 29

8. The number of threads per inch of a 3/8" screw, standard American National Coarse Threads, is MOST NEARLY
   A. 12 B. 14 C. 16 D. 18

9. To remove a drill from the sleeve or socket on a drill press one would use a
   A. drift punch
   B. drill rod
   C. drill drift
   D. chuck key

10. A cape chisel is USUALLY used for chipping
    A. grooves and slots in metal surfaces
    B. corners and concave surfaces
    C. flat surfaces or cutting through thin metal
    D. V-shape grooves on soft metals

11. An expanding mandrel is used for
    A. holding and driving cutting tools
    B. enlarging small holes
    C. extracting broken screws
    D. holding work to be machined

12. The distance from the outside diameter of a spur gear to the pitch circle of the gear is called the
    A. Dedendum   B. Total depth   C. Addendum
    D. Working depth

13. The type of gearing that eliminates the use of thrust bearings on the shafts is called
    A. Spur gearing   B. Bevel gearing
    C. Herringbone gearing   D. Spiral gearing

14. The vernier scale on a one inch micrometer is located on the
    A. Frame   B. Spindle   C. Thimble   D. Sleeve

15. A thread micrometer is used on threads for measuring
    A. minor diameter   B. major diameter
    C. pitch diameter   D. shoulder diameter

16. A sine bar is used to measure
    A. hole diameters   B. pitch diameters
    C. tapers   D. out of round surfaces

17. Which one of the following is generally used for setting a surface gauge?
    A. Gage Blocks   B. Planer Gauge
    C. Inside caliper   D. Combination square

18. A center gauge is used to check
    A. Diameters of holes   B. Threads
    C. lines on rough work   D. Tapers

19. The tool used to locate the centers on the end of a round piece of stock for turning on a lathe is called
    A. Hermaphrodite calipers   B. Outside calipers
    C. Inside calipers   D. Dividers

Question 20 refers to the sketch shown below.

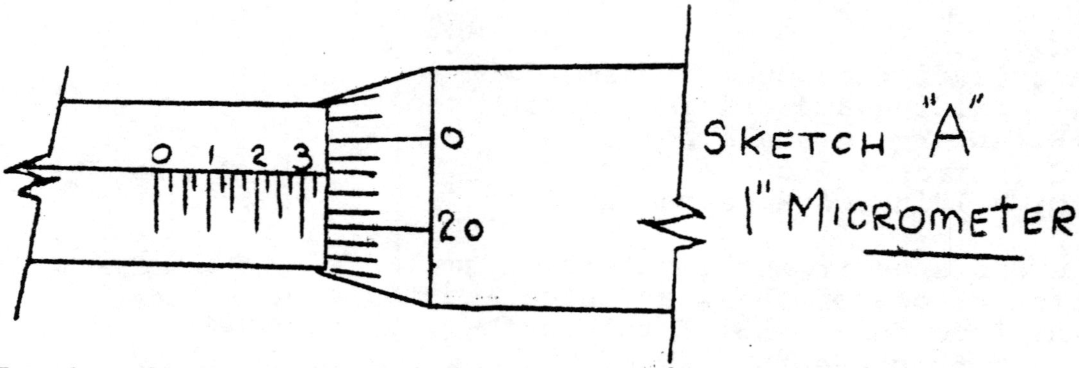

20. In the above sketch, "A", the reading shown is
A. .327 B. .323 C. .343 D. .348

Question 21 refers to the sketch shown below.

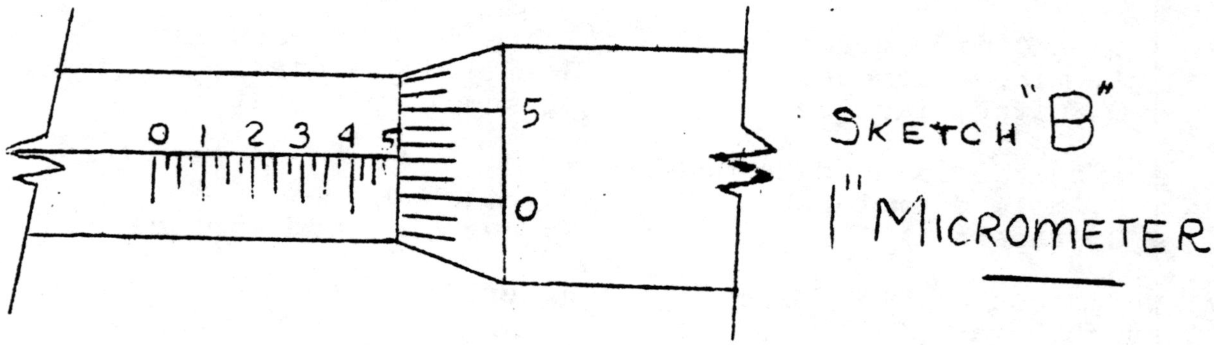

21. In the above sketch, "B", the reading shown is
A. .498 B. .462 C. .502 D. .505

22. A Vernier protractor is used in a machine shop to measure
A. Angles B. Threads C. Fillets D. Radii

23. Which one of the following would the able machinist use to check a piece of work in a lathe for concentricity?
A. Vernier Calipers B. Snap Gage
C. Dial Indicator D. Plug Gage

24. A brazed joint is generally made with the use of a
A. copper soldering iron and soft solder
B. copper soldering iron and hard solder
C. torch and hard solder
D. torch and soft solder

25. Which one of the following statements relative to the care of grinding wheels is CORRECT?
A. Wheels should be stored in a damp place.
B. Wheels should be mounted without flanges.
C. In mounting wheels on spindles, the wheel nuts should be screwed on very tightly.
D. Wheels should be equipped with blotting paper gaskets on each side.

26. Which one of the following is NOT a safe practice in a machine shop?
    A. Wearing short sleeve shirts.
    B. Wearing goggles for grinding operations.
    C. Stopping a machine for another operator in case of an accident.
    D. Slowing down moving machine parts with one's hands after the power has been shut off.

27. In grinding a piece of steel, one observes long club-shaped sparks, smooth light lines, light yellow in color, and no small stars. The steel is MOST likely
    A. high carbon steel   B. low carbon steel
    C. chrome steel   D. high speed steel

Questions 28-32.

DIRECTIONS: Column I below lists words commonly used in a machine shop and Column II gives definitions for these words. For each word listed in Column I, enter in the appropriate space on your answer sheet the capital letter in front of its definition in Column II.

COLUMN I	COLUMN II
28. Back lash	A. Used for cutting slots and edges of work.
29. Allownace	B. Used for cutting convex grooves.
30. End Mill	C. Difference in dimensions of mating parts.
31. Clearance	D. Space allowed to prevent interference.
32. Chatter	E. Vibration of tool or work.
	F. Lost motion in movable parts.
	G. Amount of variation permitted on dimensions of a machined part.

33. A machinist who makes a <u>significant</u> error makes one which is MOST NEARLY
    A. important   B. accidental   C. meaningless   D. doubtful

34. A machinist who is given <u>explicit</u> directions is given directions which are MOST NEARLY
    A. forceful   B. erroneous   C. confusing   D. definite

35. A machinist who is <u>zealous</u> in his work is one who is MOST NEARLY
    A. enthusiastic B. envious C. courteous D. patient

36. A machinist who is working on a job which is to be <u>expedited</u> is working on one which is to be MOST NEARLY
    A. supervised B. hastened C. carefully done
    D. dismantled

37. The MAIN purpose of a cutting fluid is to
    A. improve the finish B. remove the generated heat
    C. wash away the chips D. prevent rust

38. The procedure to follow in the lubrication of the machinery in a machine shop is to lubricate
    A. when you have time
    B. at regular intervals
    C. when necessary
    D. while the machinery is running

39. A steel plate is to be case hardened. This plate would MOST likely be of
    A. low carbon steel B. high carbon steel
    C. vanadium alloy steel D. high speed steel

40. Hardened steels are sometimes reheated to relieve the internal strain and to increase the toughness. This process is called
    A. Quenching B. Carburizing
    C. Tempering D. Annealing

41. If the base of a right triangle is 9" and the altitude is 12", the length of the third side will be
    A. 13" B. 14" C. 15" D. 16"

42. If a steel bar 1" in diameter and 12' long weighs 32 lbs., then the weight of a piece of this bar 5'9" long is MOST NEARLY
    A. 15.33 lbs. B. 15.26 lbs. C. 16.33 lbs. D. 15.06 lbs.

43. The diameter of a circle whose circumference is 12" is MOST NEARLY
    A. 3.82" B. 3.72" C. 3.62" D. 3.52"

44. A dimension of 39/64 inches converted to decimals is MOST NEARLY
    A. .600" B. .609" C. .607" D. .611"

45. A gear blank casting weighs 24 lbs. If the machined gear weighs 16.8 lbs. when finished, the percent of the blank which is wasted is
    A. 20 B. 30 C. 40 D. 70

46. The outside diameter of a 10 diametral pitch gear having 50 teeth is MOST NEARLY
    A. 5.2" B. 5.4" C. 10" D. 5"

47. A 9" pulley revolving at 950 R.P.M. is belted to a 20" pulley. The R.P.M. of the 20" pulley is MOST NEARLY
    A. 2100 B. 2000 C. 450 D. 427

Question 48 refers to the sketch shown below.

48. In the sketch below, the taper, per foot, is equal to
A. 3/8" B. 1 7/8" C. 2 3/8" D. 2 1/4"

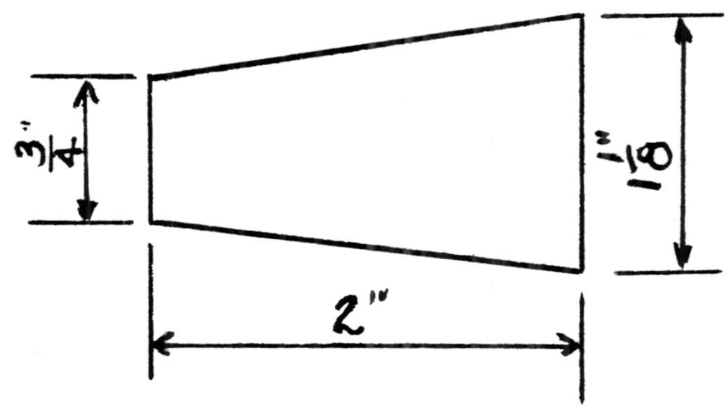

49. A grinding wheel 6" in diameter rotates at 3200 R.P.M. The surface speed, in feet per minute, is MOST NEARLY
A. 4500 B. 5500 C. 5000 D. 6000

Question 50 refers to the sketch shown below.

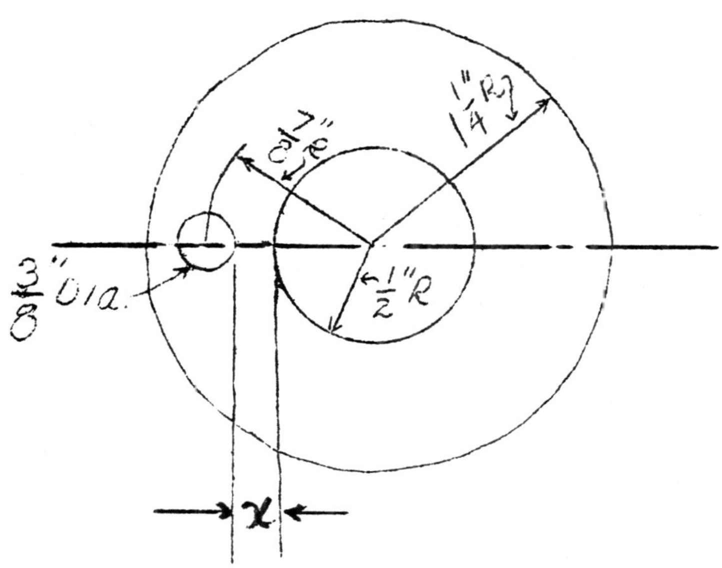

50. In the above sketch the distance "X" is equal to
A. 1/8" B. 3/16" C. 1/4" D. 5/16"

KEY (CORRECT ANSWERS)

1.	D	11.	D	21.	C	31.	D	41.	C
2.	B	12.	C	22.	A	32.	E	42.	A
3.	A	13.	C	23.	C	33.	A	43.	A
4.	C	14.	D	24.	C	34.	D	44.	B
5.	C	15.	C	25.	D	35.	A	45.	B
6.	C	16.	C	26.	D	36.	B	46.	A
7.	D	17.	D	27.	B	37.	B	47.	D
8.	C	18.	B	28.	F	38.	B	48.	D
9.	C	19.	A	29.	C	39.	A	49.	C
10.	A	20.	D	30.	A	40.	C	50.	B

EXAMINATION SECTION

TEST 1

DIRECTIONS: Each question or incomplete statement is followed by several suggested answers or completions. Select the one that *BEST* answers the question or completes the statement. *PRINT THE LETTER OF THE CORRECT ANSWER IN THE SPACE AT THE RIGHT.*

1. In machine tapping a through hole in a piece of 1/2-inch flat stock, *which* of the following taps would be *LEAST LIKELY* to be used? 1.___
   A. Bottoming B. Plug C. Taper D. Nut

2. The dimensions of a fillet should *generally* be checked with a(n) 2.___
   A. angle gage B. compass
   C. radius gage D. ring gage

3. Bronze is an alloy *PRIMARILY* composed of 3.___
   A. copper and tin B. zinc and tin
   C. copper and lead D. cadmium and lead

4. The type of milling cutter that has teeth *only* on its circumference is known as a(n) 4.___
   A. end milling cutter B. side milling cutter
   C. face milling cutter D. plain milling cutter

5. In using a plain horizontal milling machine for milling hexagons on the ends of pieces of round stock, *which* of the following set-ups produces the *FASTEST* operation? A 5.___
   A. dividing head and straddle mills
   B. rotary table and a face mill
   C. rotary table and a butt mill
   D. dividing head and an end mill

6. The lip clearance angle of a general purpose twist drill should be maintained *between* 6.___
   A. 2 and 5 degrees B. 7 and 10 degrees
   C. 12 and 15 degrees D. 17 and 20 degrees

7. Commercial tap drills are those size drills which have been adopted by the tap and die manufacturers for American Standard Form threads. These drills allow *approximately* 7.___
   A. 50% of a full thread B. 60% of a full thread
   C. 75% of a full thread D. 90% of a full thread

8. *Which* of the following reamers is *generally* recommended for machine reaming when jigs with hardened bushings are used? 8.___
   A. Adjustable blade B. Jobber's
   C. Fluted D. Rose

9. Seasoning of castings is done in order to 9.___
   A. equalize the depth of case hardening in the casting
   B. insure consistent hardness throughout the casting
   C. relieve internal stresses in the casting
   D. maintain a consistent level of carbon content in the casting

10. *Which* of the following steels is ued to make drill rod? 10.___
   A. Low carbon steel   B. Medium carbon steel
   C. High carbon steel   D. High speed steel

11. A tool used to remove a broken tap from the hole in which it was broken is known as a(n) 11.___
   A. tap extractor   B. easy -out
   C. tap puller   D. tap sinker

12. A hardie hole and a pritchel hole are *usually* found in a(n) 12.___
   A. engine block casting   B. anvil
   C. clapper box   D. compound rest

13. *Which* of the following should be used for filing lead or babbitt? 13.___
   A. Crossing file   B. Pillar file
   C. Mill file   D. Vixen file

14. The *MAIN* purpose of a file card is to 14.___
   A. determine the cut of a file
   B. clean a file which is pinned
   C. compare the coarseness of files
   D. clean a work surface prior to filing

15. A *general* rule regarding the material to be used for a lap is that the lap should be of 15.___
   A. the same material as the work being lapped
   B. a material which is harder than the work being lapped
   C. a material which is softer than the work being lapped
   D. a material which is equal in hardness to the work being lapped

16. A segment grinding wheel is *usually* used on a 16.___
   A. horizontal rotary grinder
   B. vertical spindle rotary grinder
   C. cam grinder
   D. centerless grinder

17. Magnetic chucks are *most commonly* used to hold workpieces on *which* of the following machine tools? 17.___
   A. Drill presses   B. Milling machines
   C. Surface grinders   D. Engine lathes

18. In doing internal grinding, the general practice is to use a coolant when grinding *which one* of the following materials? 18.___
   A. Bronze   B. Steel   C. Cast iron   D. Brass

19. A lathe has been used to cut a taper by the offset method. In order to turn a straight piece of work, it is necessary to *readjust* the 19.___
   A. carriage   B. cross-slide
   C. tailstock   D. headstock

20. *Which* of the following is *one* of the *MAIN* uses of the center gage? 20.___
   A. Establishing the center of a square work piece
   B. Checking and setting 60-degree threading tools
   C. Spotting the center of a piece of round stock which is to be turned on a lathe
   D. Lining up the centers of an engine lathe

KEY (CORRECT ANSWERS)

1.	A	11.	A
2.	C	12.	B
3.	A	13.	D
4.	D	14.	B
5.	A	15.	C
6.	C	16.	B
7.	C	17.	C
8.	D	18.	B
9.	C	19.	C
10.	C	20.	B

TEST 2

DIRECTIONS: Each question or incomplete statement is followed by several suggested answers or completions. Select the one that *BEST* answers the question or completes the statement. *PRINT THE LETTER OF THE CORRECT ANSWER IN THE SPACE AT THE RIGHT.*

1. If the pitch of a triple entry thread is 1/4", the lead of the thread is 1.___
   A. 1/4" B. 1/2" C. 3/4" D. 1"

2. The term "circular pitch" applies to 2.___
   A. thread cutting
   B. spline cutting
   C. trepanning
   D. gear cutting

3. A lathe dog is *properly* used with a 3.___
   A. collet
   B. faceplate
   C. four jaw chuck
   D. combination chuck

4. The *PRIMARY* reason for using multiple-thread screw-threads is that they 4.___
   A. are more easily cut than single-thread screws
   B. are stronger than single-thread screws of the same pitch
   C. have a faster travel than single-thread screws
   D. have less tendency to loosen than single-thread screws

5. The *purpose* of tempering steel after hardening is to 5.___
   A. case harden it and disperse its carbon deposits
   B. reduce its brittleness and case harden it
   C. remove internal strains and disperse its carbon deposits
   D. reduce its brittleness and remove internal strains

6. Box-type drill jigs are *most commonly* used when 6.___
   A. holes are to be drilled parallel and in the same plane
   B. one hole is to be drilled on a number of identical work pieces
   C. holes are to be drilled on different sides of the work-piece
   D. all holes are to be drilled are equally spaced

7. A dividing head is *PRIMARILY* used on a milling machine to 7.___
   A. control the ratio between the longitudinal and transverse feeds
   B. vary the angle of the cutter in relation to the work-piece
   C. adjust the cutter speed
   D. rotate the workpiece on its axis

8. Milling machine fixtures are used to perform *all* of the following functions *EXCEPT* 8.___
   A. adopting a standard milling machine to special milling operations
   B. locating the workpiece in relation to the cutter
   C. holding the workpiece against the cutting load
   D. providing gaging or setting surfaces on the workpiece

9. In testing a grinding wheel by tapping it, *which* of the following would indicate that the wheel is solid? A 9.____
   A. dull ringing sound B. dull thudding sound
   C. sharp ringing sound D. sharp vibrating sound

10. The *one* of the following machines that *usually* transmits power through a rack and pinion is the 10.____
   A. press brake B. pneumatic hammer
   C. hydraulic press D. hand arbor press

11. A shear-pin hub has a torque rating of 6,310 pound-inches when using a shear pin with a 3/16-inch neck diameter. If a shear-pin with a 3/8-inch neck diameter is substituted, the torque rating in pound-inches should be, *most nearly*, 11.____
   A. 9,500 B. 12,700 C. 18,900 D. 25,300

12. Of the following tools, the *one* which is designed to enlarge a roughly formed hole in a casting and to smooth the bore is the 12.____
   A. broach B. core drill
   C. counterboring tool D. shell reamer

13. A simple gear train consists of a driving gear with 40 teeth and a follower gear with 20 teeth. What would be the effect on the follower gear if two idler gears, one with 60 teeth and the other with 30 teeth, were placed *between* the driver and the follower gears? The 13.____
   A. speed of the follower will increase in relation to that of the driver
   B. speed of the follower will decrease in relation to that of the driver
   C. direction of rotation of the follower will be opposite to that of the driver
   D. direction of rotation of the follower will be the same as that of the driver

14. The *most commonly* used taper for standard taper-shank drills is the 14.____
   A. Morse taper B. Brown and Sharpe taper
   C. Jarno taper D. American Standard taper

15. Gears which transmit motion between shafts whose center lines intersect are *generally* known as 15.____
   A. herringbone gears B. spur gears
   C. helical gears D. bevel gears

16. The range of numbered twist drills includes drills numbered from 1 *to* 16.____
   A. 40 B. 60 C. 80 D. 100

17. A universal knee and column type milling machine *differs* from a plain knee and column type milling machine in that the universal type has a(n) 17.____
   A. adjustable overarm B. vertical knee feed
   C. swivel table D. adjustable spindle

18. Negative back-rake on a lathe cutting tool means that the face of the tool is ground so that it slopes 18.___
   A. *upward* from the point to the back
   B. *downward* from the point to the back
   C. *upward* from the cutting edge to the side
   D. *downward* from the cutting edge to the side

19. The thread dial-indicator on a lathe is *usually* found on the 19.___
   A. tailstock
   B. carriage
   C. headstock
   D. cross-slide

20. In the United States, the nominal swing of an engine lathe is *approximately* the 20.___
   A. maximum diameter which can be rotated over the ways
   B. maximum diameter which can be rotated over the cross-slide
   C. height of the centers above the ways
   D. height of the centers above the cross-slide

KEY (CORRECT ANSWERS)

1.	C	11.	D
2.	D	12.	B
3.	B	13.	C
4.	C	14.	A
5.	D	15.	D
6.	C	16.	C
7.	D	17.	C
8.	A	18.	A
9.	A	19.	B
10.	D	20.	A

TEST 3

DIRECTIONS: Each question or incomplete statement is followed by several suggested answers or completions. Select the one that *BEST* answers the question or completes the statement. *PRINT THE LETTER OF THE CORRECT ANSWER IN THE SPACE AT THE RIGHT.*

1. The length of a file is measured *from* the point *to* the end of the 1.____
   A. handle B. teeth C. tang D. heel

2. Of the following, the *MOST* sensitive drill press is the *one* in which the drill feed is controlled by 2.____
   A. hand
   B. motor speed
   C. a cam reduction gear
   D. a lead screw

3. Cape chisels are *most commonly* used to cut 3.____
   A. narrow grooves
   B. V-shaped grooves
   C. filleted corners
   D. wide flat surfaces

4. The shear angle of a milling cutter is *also* called the 4.____
   A. tooth angle
   B. clearance angle
   C. rake angle
   D. helix angle

5. The *recommended* limit of expansion of a 3/4-inch diameter expansion hand reamer is 5.____
   A. 0.003" B. 0.008" C. 0.013" D. 0.018"

6. A machine in which the work and table move back and forth under the tool is the 6.____
   A. lathe
   B. drill press
   C. planer
   D. shaper

7. A radial ball bearing which is installed on a shaft whose diameter exceeds the manufacturer's specification will *most likely* fail in a short time because the 7.____
   A. inner race cannot rotate on the shaft
   B. bearing's internal clearance will be reduced
   C. shaft will be overstressed
   D. outer race cannot rotate in the housing

8. The *MAIN* reason for using a flexible coupling to connect the shaft of a pump to the shaft of an electric motor is to 8.____
   A. allow for imperfect alignment of the two shafts
   B. reduce the starting load on the motor
   C. prevent damage to the pump impeller if it becomes obstructed
   D. connect two shafts of different diameters

9. Hermaphrodite calipers are *most commonly* used for 9.____
   A. scribing circles on flat stock
   B. locating centers of round stock
   C. checking outside diameters of round stock
   D. checking inside diameters of bored holes

10. On a one-inch micrometer caliper, fifteen exposed graduations on the barrel and twelve graduations on the thimble indicate a reading of 10.____
   A. 0.612 inches B. 0.537 inches
   C. 0.462 inches D. 0.387 inches

11. *Which* of the following tools is *most commonly* used for leveling and lining up work? 11.____
   A. Vernier caliper B. Center gage
   C. Surface gage D. Trammel points

12. A micrometer has a range of one inch, by thousandths. The *number* of threads per inch on the spindle of this micrometer is 12.____
   A. 25 threads per inch B. 40 threads per inch
   C. 50 threads per inch D. 80 threads per inch

13. The *one* of the following instruments that is used in a machine shop to measure angles is a 13.____
   A. scriber B. caliper C. micrometer D. protractor

14. A telescoping gage is used in measuring 14.____
   A. inside diameters B. extremely long lengths
   C. outside diameters D. depths of holes

15. *Which* of the following instruments is *usually* used to measure surface roughness (rms) of a finished metal surface? 15.____
   A. Profilometer B. Galvanometer
   C. Comparator D. Reflectoscope

16. *Which* of the following measuring tools is *most commonly* used in conjunction with a small-hole gage? A(n) 16.____
   A. micrometer depth gage B. inside caliper
   C. dial indicator D. outside micrometer

17. When layihn tout six equally-spaced holes on a 6"-diameter circle, the dividers should be set *to* 17.____
   A. 2.750" B. 3.000" C. 3.240" D. 3.500"

18. When facing on a lathe, using the compound rest which is set at a 30-degree angle from the face of the work, feeding the tool 0.010" on the dial of the compound rest will result in a cut of, *most nearly*, 18.____
   A. 0.0033" B. 0.0050" C. 0.0067" D. 0.0100"

19. A table of feeds and speeds indicates that the proper cutting speed for a proposed lathe cut is 80 fpm. If the workpiece is cylindrical and 5 inches in diameter, the speed to use for this work is, *most nearly*, 19.____
   A. 49 rpm B. 61 rpm C. 76 rpm D. 154 rpm

20. 20.___

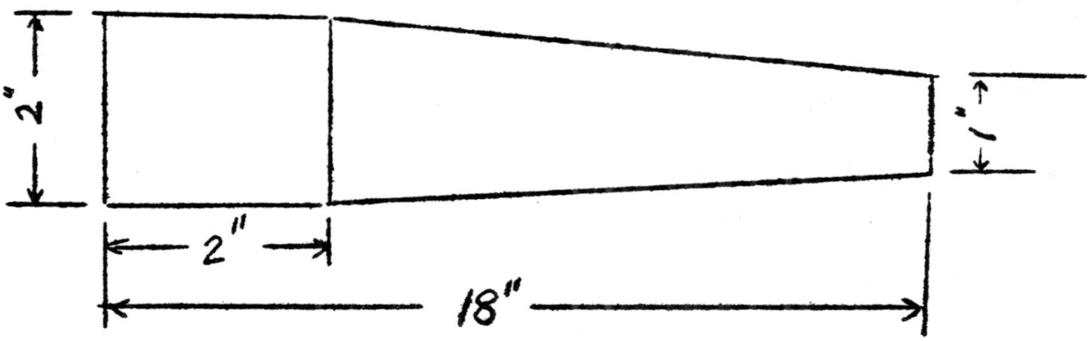

NOTE: Above sketch is not to scale.

In the above sketch, the taper per foot of the tapered section is

A. .500" B. .625" C. .750" D. .875"

KEY (CORRECT ANSWERS)

1.	D	11.	C
2.	A	12.	B
3.	A	13.	D
4.	D	14.	A
5.	B	15.	A
6.	C	16.	D
7.	B	17.	B
8.	A	18.	B
9.	B	19.	B
10.	D	20.	C

TEST 4

DIRECTIONS: Each question or incomplete statement is followed by several suggested answers or completions. Select the one that *BEST* answers the question or completes the statement. *PRINT THE LETTER OF THE CORRECT ANSWER IN THE SPACE AT THE RIGHT.*

Questions 1 - 5.

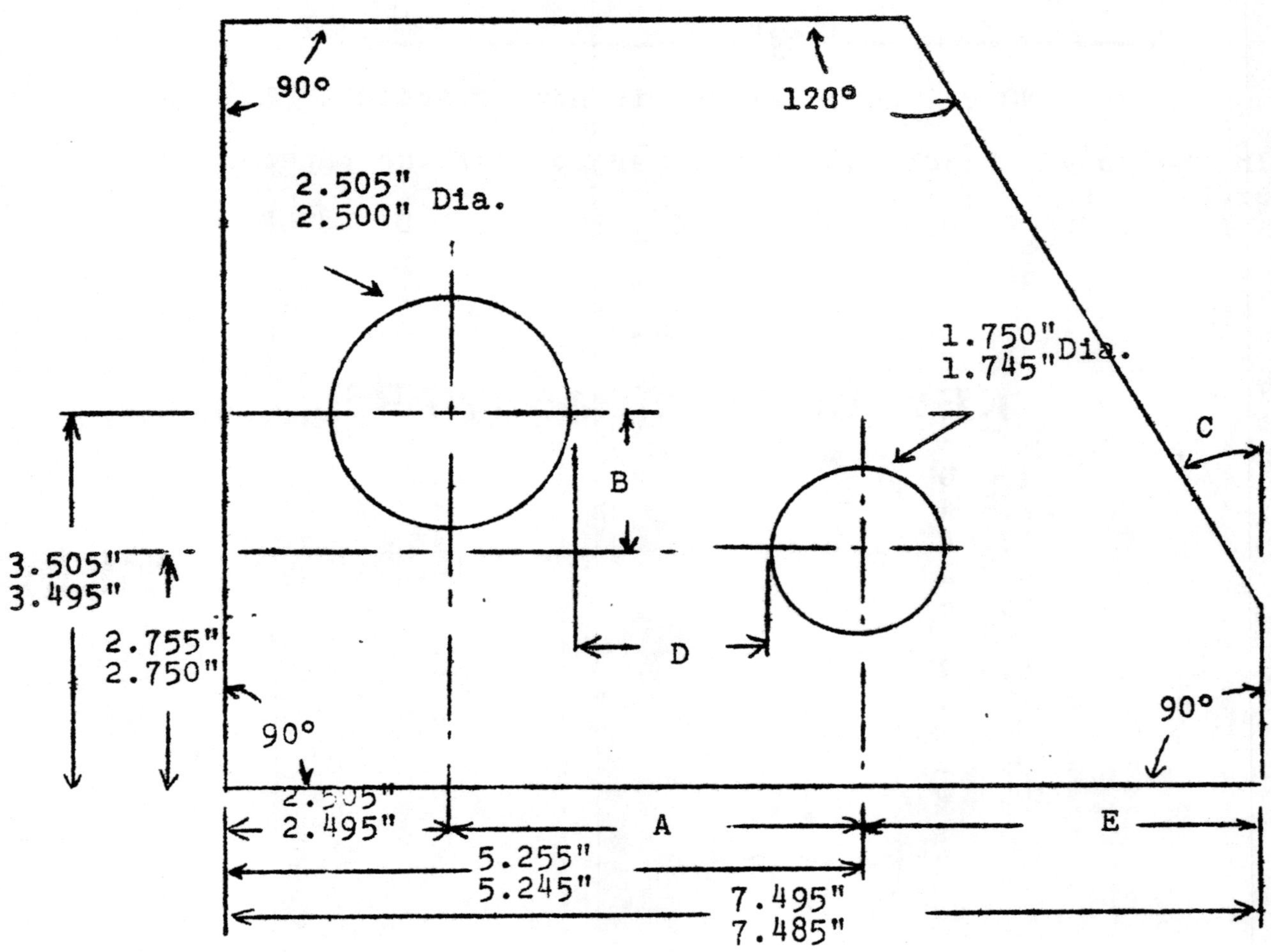

NOTE: Not To Scale

DIRECTIONS: Questions 1 to 5 inclusive are to be answered *only* in accordance with the above sketch.

1. The *MINIMUM* value of dimension A is 1.___
   A. 2.740" B. 2.745" C. 2.750" D. 2.760"

2. The *MAXIMUM* value of dimension B is 2.___
   A. .740" B. .745" C. .750" D. .755"

3. Angle C is 3.___
   A. 15 degrees B. 30 degrees C. 45 degrees D. 60 degrees

4. The *MAXIMUM* possible variation in dimension D is 4.___
   A. .005" B. .015" C. .025" D. .030"

5. The *MAXIMUM* possible variation in dimension E is 5.___
   A. .025" B. .020" D. .015" D. .010"

6. 6.___

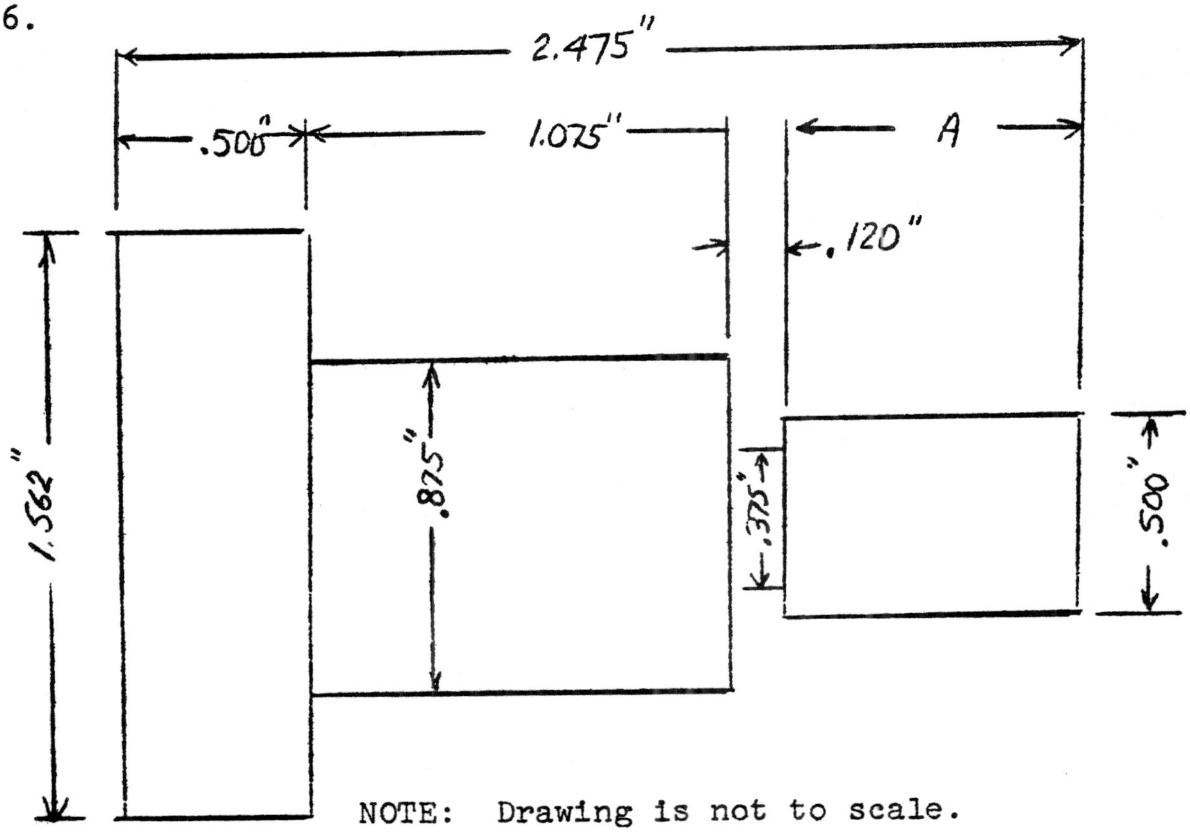

NOTE: Drawing is not to scale.

In the above drawing, the dimension indicated by the letter "A" is
A. .900" B. .840" C. .780" D. .660"

7. A string of dimensions on a shop drawing are 2 1/2", 8 3/8", 4 1/4" and 6 13/16". The overall dimension should be 7.___
A. 20 15/16" B. 21 9/16" C. 21 15/16" D. 22 15/16"

8. The diameter of a shaft was measured and the following readings were obtained: 2.874", 2.869", 2.872", 2.873" and 2.874". The average of these five readings is, *most nearly*, 8.___
A. 2.871" B. 2.872" C. 2.873" D. 2.874"

Questions 9 - 12.

DIRECTIONS: Questions 9 to 12 inclusive are to be answered *only* in accordance with the following paragraph.

Cylindrical surfaces are the most common form of finished surfaces found on machine parts, although flat surfaces are also very common; hence, many metal-cutting *PROCESSES* are for the purpose of producing either cylindrical or flat surfaces. The machines used for cylindrical or flat shapes may be, and often are, utilized also for forming the various irregular or special shapes required on many machine parts. Because of the prevalence of cylindrical and flat surfaces, the student of manufacturing practice should learn first about the machines and methods employed to produce these surfaces. The cylindrical surfaces may be internal as in holes and cylinders.

Any one part may, of course, have cylindrical sections of different diameters and lengths and include flat ends or shoulders and frequently there is a threaded part or possibly some finished surface that is not circular in cross-section. The prevalence of cylindrical surfaces on machine parts explains why lathes are found in all machine shops. It is important to understand the various uses of the lathe because many of the operations are the same fundamentally as those performed on other types of machine tools.

9. According to the above paragraph, the *most common* form of finished surfaces found on machine parts is 9.___
   A. cylindrical B. elliptical
   C. flat D. square

10. According to the above paragraph, *any one* part of cylindrical surfaces may have 10.___
   A. chases B. shoulders C. keyways D. splines

11. According to the above paragraph, lathes are found in all machine shops because cylindrical surfaces on machine parts are 11.___
   A. scarce B. internal C. common D. external

12. As used in the above paragraph, the word *processes* means 12.___
   A. operations B. purposes
   C. devices D. tools

Questions 13 - 14.

DIRECTIONS: Questions 13 and 14 are to be answered *only* in accordance with the following paragraph.

The principle of interchangeability requires manufacture to such specification that component parts of a device may be selected at random and assembled to fit and operate satisfactorily. Interchangeable manufacture, therefore, requires that parts be made to definite limits of error, and to fit gages instead of mating parts. Interchangeability does not necessarily involve a high degree of precision; stove lids, for example, are interchangeable but are not particularly accurate, and carriage bolts and nuts are not precision products but are completely interchangeable. Interchangeability may be employed in unit-production as well as mass-production systems of manufacture.

13. According to the above paragraph, in order for parts to be interchangeable, they must be 13.___
   A. precision-machined B. selectively-assembled
   C. mass-produced D. made to fit gages

14. According to the above paragraph, carriage bolts are interchangeable because they are 14.___
   A. precision-made
   B. sized to specific tolerances
   C. individually matched products
   D. produced in small units

Questions 15 - 16.

DIRECTIONS: Questions 15 and 16 are to be answered *only* in accordance with the following paragraph.

Metal spraying is used for many purposes. Worn bearings on shafts and spindles can be readily restored to original dimensions with any desired metal or alloy. Low-carbon steel shafts may be supplied with high-carbon steel journal surfaces, which can then be ground to size after spraying. By using babbitt wire, bearings can be lined or babbitted while rotating. Pump shafts and impellers can be coated with any desired metal to overcome wear and corrosion. Valve seats may be re-surfaced. Defective castings can be repaired by filling in blow-holes and checks. The application of metal spraying to the field of corrosion resistance is growing, although the major application in this field is in the use of sprayed zinc. Tin, lead, and aluminum have been used considerably. The process is used for structural and tank applications in the field as well as in the shop.

15. According to the above paragraph, worn bearing surfaces on shafts are metal-sprayed in order to 15.___
A. prevent corrosion of the shaft
B. fit them into larger-sized impellers
C. return them to their original sizes
D. replace worn babbitt metal

16. According to the above paragraph, rotating bearings can be metal sprayed using 16.___
A. babbitt wire B. high-carbon steel
C. low-carbon steel D. any desired metal

17. A portable fire extinguisher identified with the letter "C" within a blue circle is *usually* used for fighting small fires involving 17.___
A. ordinary combustibles such as wood, paper and textiles
B. flammable liquids and gases such as oil, gasoline, paint and grease
C. electrical wiring and equipment
D. combustible metals such as magnesium, potassium, and sodium

18. The *SAFEST* method of disposing of a small accumulation of metal chips on a machine is to 18.___
A. gather them up with gloved hands and deposit them in a scrap barrel
B. sweep them into a pan with a brush
C. use a fork to deposit them in a barrel
D. blow them off with an air gun

19. The American Standard Safety Color Code for Marking Physical Hazards recommends that the color orange be used to 19.___
A. designate the location of first-aid and safety equipment
B. identify radiation hazards
C. designate housekeeping and traffic markers
D. highlight hazardous parts of machines or electrical equipment

20. The *most important* reason for using a three-pronged plug on a portable electric drill is to 20.___
   A. avoid overloading the electric circuit
   B. allow the use of either AC or DC
   C. ground the drill casing to reduce the risk of accidental shock
   D. make the drill safe to use when standing on wet floors

KEY (CORRECT ANSWERS)

1.	A	11.	C
2.	D	12.	A
3.	B	13.	D
4.	C	14.	B
5.	B	15.	C
6.	C	16.	A
7.	C	17.	C
8.	B	18.	B
9.	A	19.	D
10.	B	20.	C

SUPERVISION, ADMINISTRATION & MANAGEMENT

EXAMINATION SECTION

TEST 1

DIRECTIONS: Each question or incomplete statement is followed by several suggested answers or completions. Select the one that BEST answers the question or completes the statement. *PRINT THE LETTER OF THE CORRECT ANSWER IN THE SPACE AT THE RIGHT.*

1. The average foreman probably spends the GREATEST amount of time on the job in 1. ...
   A. getting the proper tools, equipment and supplies for his men
   B. supervising his men and talking to them about the work
   C. preparing and writing reports and doing other paperwork
   D. reading rules, regulations and orders which have to be followed by the foreman and work crew

2. The MAIN reason for requiring an employee to call the office in the morning if he is sick and can't come to work is because the foreman 2. ...
   A. may have to make a different plan for the man's work
   B. must be sure that the employee is not giving a "phony" excuse
   C. must report unplanned absences to his own supervisor right away
   D. must keep a record of all absences

3. Of the following, the MOST common cause of accidents on the job is 3. ...
   A. cluttered job sites
   B. improper and faulty material
   C. carelessness of employees
   D. poorly maintained equipment

4. In training workers in safety, it is LEAST important for the foreman to impress on them that they must 4. ...
   A. correct or report unsafe conditions quickly
   B. keep tools and equipment clean, orderly, and in good working order
   C. follow safety instructions carefully
   D. understand instructions completely the first time they are given

5. Many supervisors make rounds of their men at work, talking to one person at a time about safety on the job. When approaching a worker at such time, it is a GOOD idea for the foreman to 5. ...
   A. watch the man until he does something wrong and then talk to him about that
   B. walk right up to the man and deliver a lecture on safety in general
   C. compliment the man for some proper safety action he has taken
   D. hand the man a copy of "Rules for Safety on the Job"

6. It is NOT a basic responsibility of the foreman to make sure that employees 6. ...
   A. use the right tools and equipment for the job
   B. report all injuries and get first aid promptly
   C. drive in to work carefully
   D. use required protective equipment on the job

7. To make sure that a new worker understands and is following safe methods of work, it would be BEST for the foreman to 7. ....
   A. question him about the rules of safety
   B. instruct him in safety and give him close supervision on the job
   C. assign him only to work on which he has experience
   D. have an experienced fellow-worker, who knows all the rules of safety, break him in

8. Although the employees under your supervision know the safety precautions they should take in doing a job, accidents that happen will MOST often be due to 8. ...
   A. a tendency to forget these precautions
   B. carelessness of the supervisor
   C. deliberate attempts by an employee to injure a fellow-worker
   D. misunderstanding of the precautions on the part of the employee

9. One of your men was in an accident on the job.A few days later a man who says he is a newspaper reporter comes and questions you about what happened. You should tell him 9. ...
   A. you can't give him any information because he is not a city employee
   B. you did not see what happened but that he can talk to other members of the crew who saw the accident
   C. to go to the Central Office of your Department for official information about the accident
   D. the accident was not serious

10. A foreman who writes a report should NOT use 10. ...
    A. everyday language B. poor spelling
    C. any short sentences D. any long words

11. A foreman who is responsible for storing and issuing supplies should keep a record of supplies received, how much is used, and what is on hand in the storeroom. The BEST reason for keeping such a record is to 11. ...
    A. determine at any time the cost of supplies
    B. check up on the accuracy of the centrak purchasing and delivery sections
    C. know when to order additional supplies
    D. let employees know that it is easy to find out if supplies are being wasted

12. An employee asks you a question about the Department's Time and Leave Regulations. You do not know the answer to the employee's question. In this situation, which one of the following statements would be BEST for you to use? 12. ...
    A. "I don't know the answer, but I'll find out for you."
    B. "How should I know? Nobody tells me anything."
    C. "The answer is simple. Just read the Time and Leave Regulations given to you."
    D. "That's a good question. Call Central Office and let me know the answer they give you."

13. While supervising a work crew, a foreman notices an employee doing some work incorrectly. It is a minor thing and the foreman decides to wait until he has more things to discuss with the employee. The action of the foreman in this situation is 13. ...
    A. *correct;* because the mistake was minor and the worker would think that the foreman is picking on him if he takes it up now
    B. *wrong;* the foreman should take the matter up with all the employees as they can all learn from it
    C. *correct;* because the foreman would be wasting his time and the time of the worker if he interrupts to take up a minor matter
    D. *wrong;* the mistake should be taken up now while it is fresh in the mind of the worker and the foreman

14. A new worker MOST often has difficulty learning the work because he 14. ...
    A. doesn't want to pay for the tools and materials that he may spoil
    B. feels that as soon as he has learned the job he will be expected to do as much work as older and more experienced workers
    C. is nervous and has no confidence in himself
    D. wants to learn the job his own way

15. A new method of doing a certain job usually assigned to your work crew is to be introduced by the Department. In order to get the crew ready to learn this new method, it would be BEST for you, as foreman, to 15. ...
    A. warn the crew that the men who don't learn fast may be dismissed
    B. explain to the crew why the new method is better than the old method
    C. say nothing, but post a notice on the bulletin board that training sessions will take place very soon
    D. caution the men that learning new methods is more difficult for the older men than the younger men

16. A foreman should discuss an employee's annual service rating with him 16. ...
    A. privately in an office or other suitable place
    B. during a group meeting with the work crew so that good and poor employees can be compared at the same time
    C. when the foreman's supervisor is present so that he can back up the foreman

D. only if the rating is unsatisfactory and disciplinary action is recommended

17. You have noticed that one of your workers who recently finished his probationary period has been absent last month on four Mondays in a row and has also signed out for ½ day on pay day after getting his check. In this situation, you should 17. ..
   A. call in the employee and discuss this pattern of absences with him
   B. refer the employee to the Central Personnel Office as this is an indication that the employee is an alcoholic
   C. take no action at this time but plan to discuss abuses of the time and leave regulations at the next general meeting with your employees
   D. discuss the matter with your supervisor before you do anything

18. Suppose that one of your newer men is not doing satisfactory work. Of the following, the BEST thing for you to do first is to 18. ..
   A. find an easier assignment for the employee
   B. discuss with your supervisor the possibility of transferring this employee to another work crew
   C. recommend that the employee be dismissed
   D. find out why the employee is doing poor work

19. A supervisor should treat each member of his work crew as an individual. 19. ..
This practice is USUALLY
   A. *good* because it shows the crew that the supervisor is interested in each man as a person
   B. *bad* because a supervisor should not show any favoritism
   C. *good* because it will help the supervisor to become well liked by his crew
   D. *bad* because it will cause arguments among the members of the crew

20. One of the better workers under your supervision tells you he is quitting the job. As a good foreman, you should NOT 20. ..
   A. try to find out his reason for quitting
   B. say to him "O.K.- good riddance"
   C. ask him what he thinks of the department of your unit
   D. ask him if he thinks working conditions can be improved

21. Some of your employees come to you very much upset and tell you that they have heard through the "grapevine" that there are going to be some layoffs. In this situation, it would be BEST for you to tell them 21. ..
   A. you will check into this and let them know if there is anything to it
   B. there is no truth to the rumor
   C. to write to Central Office to complain about the coming layoff
   D. they should not pay attention to "grapevine" rumors as they are usually wrong

22. Of the following actions, the one that would be LEAST likely to help the foreman get and keep the respect of his employees is 22. ...
   A. keeping the promises he makes as a supervisor
   B. taking their part when he believes their grievances justified
   C. listening to their complaints about conditions on the job and doing nothing about the complaints
   D. giving them credit where credit is due

23. One way in which GOOD morale can be developed and maintained by a foreman is to 23. ...
   A. assume all the employees are equal and refuse to recognize special ability
   B. disregard suggestions from employees and take an "all-wise" attitude toward them
   C. tell the employees about his personal problems, but do not discuss their troubles with them
   D. be direct and positive in giving instructions but soften his criticism, if possible, when making corrections

24. Many of a supervisor's problems remain unsolved because of a negative attitude. 24. ...
   Which one of the following statements by a foreman shows the most POSITIVE attitude toward solving a problem?
   A. "We tried that before and it didn't work."
   B. "It's different, but let's give it a try."
   C. "It isn't practical for operating people."
   D. "Regulations will stop us from getting it approved."

25. For the past few days one of your best workers has been doing poor work and coming in late. He explains to you that he is worried about illness in his family. When talking to him, it would be BEST for you, as foreman, to 25. ...
   A. tell him not to let his personal problems interfere with his work
   B. advise him to see a doctor for help in getting a good night's sleep
   C. ask him if he would like to take some time off to be with his family
   D. tell him his excuse is not acceptable because all men have family troubles at times

KEY (CORRECT ANSWERS)

1. B
2. A
3. C
4. D
5. C

6. C
7. B
8. A
9. C
10. B

11. C
12. A
13. D
14. C
15. B

16. A
17. A
18. D
19. A
20. B

21. A
22. C
23. D
24. B
25. C

TEST 2

DIRECTIONS: Each question or incomplete statement is followed by several suggested answers or completions. Select the one that BEST answers the question or completes the statement. *PRINT THE LETTER OF THE CORRECT ANSWER IN THE SPACE AT THE RIGHT.*

1. An immediate superior has a tendency to give his foremen verbal orders which lack sufficient detail and clarity. In this situation the BEST procedure for a foreman to follow would be to 1. ...
   A. suggest that this superior give all his orders in writing
   B. complain to the head of his department
   C. ask this superior for clarification at the time the orders are received
   D. go back to this superior repeatedly during each job for further clarification

2. If a foreman makes a serious error in the details of an order issued to his men, he should 2. ...
   A. correct his error as soon as possible
   B. immediately go to his superior and ask for his advice
   C. wait and see whether his men discover the error before doing anything about it
   D. ignore the error so as to avoid the risk of having his men think he is not good enough for his job

3. If workers continue to work effectively even when their foreman is out sick, it would MOST probably indicate that these workers 3. ...
   A. are trying to embarrass the foreman
   B. are in constant fear of the foreman
   C. have proved that the foreman is serving no useful purpose
   D. are properly trained and the foreman has their full co-operation

4. A foreman assigns a new man to work with a capable and experienced employee. After one day, the new man complains to the foreman that he knows more about the job tham the experienced employee and does not understand why he has to follow the man's directions. The BEST procedure for the foreman to follow in handling this situation is to 4. ...
   A. reprimand the experienced employee for not knowing his job
   B. have the new man work by himself and, if he makes a mistake, reprimand him
   C. tell the new man that the experienced worker is capable and his directions must be followed
   D. have a heart-to-heart talk with the experienced worker and ask him to go along with the new man

5. Of the following the one which is NOT a responsibility of a foreman is 5. ...
   A. improving work methods
   B. keeping his men satisfied
   C. settling disputes among fellow foreman
   D. suggesting better safety practices to higher supervision

6. A foreman can BEST maintain effective discipline by 6. ...
   A. never giving an inch
   B. watching his men constantly
   C. being firm in important matters
   D. bringing all violators up on charges

7. One of the BEST ways a foreman can cooperate with his superiors would be for him to 7. ...
   A. push his men continually
   B. constantly report all details of a job in progress
   C. never make a decision without his superior's approval
   D. accept full responsibility for the work done by the men assigned to him

8. Several workers start a discussion, during working hours, of rumored changes in working conditions. The BEST procedure for the foreman to follow would be to break up the discussion and tell the men that 8. ...
   A. they should not listen to rumors
   B. he will check on the rumor for them
   C. changes in working conditions are not their concern
   D. they should not be concerned until changes are officially announced

9. The FIRST step for a foreman to take in dealing with a worker's grievance when the complaint has little or no merit is to 9. ...
   A. process the grievance in accordance with the standard procedures of his agency
   B. acknowledge the grievance without taking any formal action
   C. discuss the grievance with the worker and point out its weaknesses
   D. warn the worker that grievances of this nature may subject him to disciplinary action

10. The MOST important thing for a foreman to bear in mind in trying to establish a good safety record is that 10. ...
   A. most accidents can be prevented by observing safety rules
   B. some accidents are unavoidable if production is to be maintained at the required levels
   C. certain individuals who are accident-prone will defeat all measures taken to protect them
   D. a certain number of accidents will happen to his men regardless of the precautions taken

11. In order to secure the respect of his men, a newly appointed foreman should 11. ...
    A. avoid giving orders until he is completely familiar with his new job
    B. readily admit to his men that he is new at his job and ask for their cooperation
    C. conceal his lack of familiarity with his new job by maintaining a "know-it-all" attitude
    D. make it clear to his men that he will discipline anyone taking advantage of him because of his lack of familiarity with his new job

12. It is a GOOD supervisory practice for a foreman to 12. ...
    A. reprimand a worker in front of other workers
    B. speak only in vague terms when criticizing a worker
    C. give the worker a chance to reply to the foreman's criticism
    D. wait until the entire job is done before criticizing a worker

13. The MAIN purpose in giving some employees instructions in first aid is to 13. ...
    A. reduce the number of accidents
    B. save money on compensation cases
    C. eliminate the need for calling doctors
    D. be able to provide emergency aid if needed

14. If a foreman must leave his men for an hour or two, he should 14. ...
    A. reassign the workers to another job until he returns
    B. clearly designate someone to be in charge during his absence
    C. call his immediate superior and ask him to send another foreman to the job site
    D. report his intended movements to his men and tell them they are "on their honor" until he returns

15. The BEST thing for a foreman to do if he finds he is disliked by his men is to 15. ...
    A. supervise them more rigidly
    B. temporarily overlook slight infractions of the rules
    C. ask his immediate superior to have a talk with the men
    D. take stock of himself and his actions to see whether he is at fault

16. A foreman's superior complains that he could not find the foreman at his post and that his men were idle during the time he was not there. The MOST important thing for this foreman to do as a result of this incident is to 16. ...
    A. improve his supervisory techniques
    B. disregard such an unreasonable complaint
    C. make up an excuse for being away from the job
    D. make certain he is never away from his assigned location again

17. In a report to his superior, a foreman is recommending the adoption of a radical change in a standard operating procedure. The MOST important information to be included in this report is 17. ...
    A. a list of the reasons for recommending the change
    B. the names of the men in his crew who favor this change
    C. the number of records to be kept if this change is adopted
    D. a complete and detailed description of the present procedure

18. One method a foreman may use to evaluate the ability of his men CORRECTLY is to 18. ...
    A. find out which men have good attendance records
    B. observe their work on various types of assignments
    C. find out what the men think of each other's ability
    D. gauge their ability by the speed with which they work

19. A foreman has checked a certain job and is writing a report to his superior concerning the job status. Of the following, the item that need NOT be included in this report is 19. ...
    A. his signature
    B. the date he wrote the report
    C. the date he made the inspection
    D. the names of the men he assigned to do the job

20. Of the following, the item which would LEAST affect the time it takes to complete a specific job is 20. ...
    A. the working conditions on the job
    B. the cost of the materials used on the job
    C. whether it is an exterior or interior job
    D. the availability of manpower at the time the job is performed

21. A foreman can MOST effectively prevent a waste of manpower by 21. ...
    A. curtailing privileges
    B. planning each job properly
    C. advancing deadlines for completing jobs
    D. requiring progress reports from his painters

22. A foreman finds that his superior, because of personal problems, cannot concentrate sufficiently to make needed decisions. The foreman should 22. ...
    A. go to higher supervision for these decisions
    B. be persistent and force his superior to make decisions
    C. make appropriate recommendations to this superior for his approval
    D. discuss the matter with the other foremen and rely on their decisions

23. If a foreman observes one of his men doing an assigned job in a way which is unsatisfactory, the foreman should 23. ...
   A. reassign the job to another man
   B. pass no comment until the job is finished
   C. make certain not to assign this type of job to him again
   D. tell the man the proper method to use before allowing him to continue

24. In order to train a new man to perform a routine job, a foreman assigns a worker to instruct him. After the customary "breaking-in" period, the foreman finds that the new man has made little progress. In this case, the foreman should 24. ...
   A. investigate to find out whether the job is too difficult for a beginner
   B. report to his immediate superior that the new man is not interested
   C. coninue the training period indefinitely until the new man is qualified
   D. find out whether the worker or the new man is responsible for the poor results

25. A citizen has complained of sloppy work being done on a public guilding near his office. The foreman in charge is told and report on the citizen's complaint. According to good supervisory practice, the one thing the foreman should NOT do is 25. ...
   A. speak to the men on the job about the complaint
   B. ignore the complaint unless the citizen complains again
   C. go to the job site himself and personally observe the work
   D. tell the citizen he will check on the complaint and take appropriate action

KEY (CORRECT ANSWERS)

1.	C	11.	B	21.	B
2.	A	12.	C	22.	C
3.	D	13.	D	23.	D
4.	C	14.	B	24.	D
5.	C	15.	D	25.	B
6.	C	16.	A		
7.	D	17.	A		
8.	B	18.	B		
9.	C	19.	D		
10.	A	20.	B		

TEST 3

DIRECTIONS: Each question or incomplete statement is followed by several suggested answers or completions. Select the one that BEST answers the question or completes the statement. *PRINT THE LETTER OF THE CORRECT ANSWER IN THE SPACE AT THE RIGHT.*

Questions 1-4.

DIRECTIONS: Questions 1 to 4, inclusive, are based on the situation described below. In answering these questions, consider only the information given in this paragraph.

A foreman is assigned with a crew of workers of his own selection to work on a large area that is subdivided into many small offices which will be manned by a skeleton staff. This job must be completed in one day. This means that the job must be finished in less time than normally would be allowed for this type of work.

1. It would be preferable for the foreman to select men for this crew who 1. ...
   A. are his personal friends
   B. have done this type of rush job before
   C. have great endurance rather than experience
   D. will constantly ask questions about the work being done

2. If one of the most experienced workers has a suggestion which he feels will expedite this job, the foreman should 2. ...
   A. consider whether the suggestion is worthwhile
   B. ignore the suggestion since considering it will only waste time
   C. send this worker to the next higher superior to ask for approval
   D. adopt the suggestion immediately since it comes from one of the best workers

3. The foreman can BEST minimize complaints from the office personnel by seeing to it that his men 3. ...
   A. report to work on time
   B. answer all questions by the office personnel in detail
   C. refrain from unnecessarily obstructing areas not being repaired
   D. ask the office personnel for suggestions on how to expedite the job

4. As one of his men completes part of his assignment, the foreman asks him to go over his work. The worker objects because he feels this will slow him down. The foreman should 4. ...
   A. assign the job to another worker
   B. threaten the worker with disciplinary action
   C. explain to the worker the reason for the request
   D. inform the worker that it is his right to take the matter up with the foreman's immediate superior

Questions 5-9.

DIRECTIONS: Questions 5 to 9, inclusive, refer to the following list of supervisory situations. In answering these questions, consider only these situations.

SUPERVISORY SITUATIONS

K. A foreman has periodic meetings with his workers to discuss any problems they might have in regard to their assignments.

L. A foreman has each worker submit periodic reports evaluating the performance of his other workers.

M. A foreman is a perfectionist and insists that his men be perfectionists.

N. A foreman tells newly appointed employees that more work is expected from them than from his more experienced workers.

O. A highly specialized job is usually assigned to a particular foreman. He has trained two of his most competent workers to do this job.

P. A foreman praises his men at every opportunity but he also criticizes them when he feels it is justified.

Q. A foreman insists on high quality work and greater output from his more competent workers.

R. A foreman compliments a worker on his meeting a deadline for a rush job. He does this orally and within hearing of the other men who also worked on this job.

S. A foreman grants special privileges to those workers who do the best work.

5. The foreman who will probably receive the LEAST amount of respect from his men is the one described in situation 5. ...
   A. K B. L C. O D. P

6. The situation MOST likely to produce a higher accident rate on the job is 6. ...
   A. N B. O C. P D. R

7. The foreman MOST likely to keep morale high among his men is the one described in both situations 7. ...
   A. R and S B. N and Q C. L and M D. K and P

8. The foreman whose actions are considered good supervisory practice is the one described in both situations 8. ...
   A. K and O B. L and Q C. N and S D. M and K

9. The action LEAST likely to stir up resentment is the one described in situation 9. ...
   A. M B. N C. P D. R

10. When acquainting a group of newly appointed workers with the nature of their duties, it would be LEAST important to include information pertaining to 10. ...
   A. holidays B. safety practices
   C. housekeeping procedures D. lunch periods

11. Upon learning that one of your subordinates went to your superior and, by a misrepresentation of facts, obtained his consent to a request you denied, your BEST course of action would be to 11. ...
   A. demand that the man be brought up on charges
   B. do nothing to avoid prolonging the incident
   C. explain the situation to your superior
   D. request that the man be transferred

12. A foreman of another department complains to you that some of your men did not cooperate fully on a job which was done jointly with his men. With respect to this complaint you should 12. ...
   A. investigate to substantiate or disprove the charge
   B. ignore the matter because the job is finished
   C. tell the other foreman that your men always cooperate
   D. simply cite instances when his men did not cooperate with you

13. Lately, one of your senior employees is getting into many arguments with the other members of the work gang. You decide to talk to him about this. Which one of the following is the BEST way to handle the situation? 13. ...
   A. Start off the interview by telling the employee that, much as you don't like to, you will have to discipline him for his actions unless he can come up with a good excuse
   B. Listen to what the employee has to say before working out a solution to the problem
   C. Criticize the employee for his behavior as he is experienced and should know his way around
   D. Cross-examine the employee's statements to show him how wrong he has been

14. An employee complains to you, the foreman, that he is getting more dirty jobs than any one else in the crew. You tell him that you will check into it. When you finish your investigation, you find that the complaint is not justified. At this point, it would be BEST for you to 14. ...
   A. forget about the complaint because there is nothing to it
   B. refer the matter to your supervisor because the employee isn't telling the truth
   C. tell the employee why the complaint isn't justified
   D. caution the employee that he is making trouble for himself by complaining too much

15. An older, experienced worker and a new worker have come to you, the foreman, to settle an argument between them concerning the work assignment on the job. Of the following, the BEST way to settle this argument and maintain good employee morale is for you to 15. ...
   A. tell the new worker that he should always follow the advice of the older worker
   B. take the side of the new worker to show the new man that the supervisor is willing to listen to new employees
   C. tell both men that they shouldn't argue about the orders and work assignments of a foreman
   D. have each man present his side of the argument and then work out an agreement with them

16. As a new foreman you are put in charge of a work crew whose men have been working together for a few months. You don't know any of the men in this crew. In making work assignments at the beginning, it would probably be BEST for you to 16. ...
   A. ask the employees for suggestions on how to change their assignments
   B. let the employees choose their own job assignments to show them that you will be fair
   C. tell the employee with the most seniority to make out the work assignments and have him take responsibility for the work
   D. continue the men on the same assignments until you get to know them and their work better

17. As a foreman you are responsible for a large work crew which is divided into two sections. Each section works at different locations and does different work. For you to assign one employee of each section to be in charge when you are not there would be 17. ...
   A. *bad* because as a foreman you are personally responsible for everything that happens to the employees under your supervision
   B. *good* because you can't directly supervise both sections at the same time
   C. *bad* because a foreman should not turn over any of his authority and responsibility to an employee under his supervision
   D. *good* because you can hold that employee responsible if anything goes wrong

18. Assume that a normally sober man appears on the job intoxicated. Of the following, the BEST procedure for a foreman to follow is to 18. ...
    A. give the man an easy job so that he cannot get hurt
    B. let the man "sleep it off" in the morning and put him to work in the afternoon
    C. let the man work at his normal duties but keep an "eye" on him
    D. send him home for the day

19. The foreman has decided to change the procedure that must be followed in making certain types of repairs. The one of the following statements concerning the new procedure that is CORRECT is: 19. ...
    A. The men should know why the procedure is being changed because they will then be more interested in the job
    B. The men do not have to know the reason for the change because they need do only the work as they are told
    C. The men shpuld know why the procedure is being changed so that they can decide which method of doing the job is better
    D. The men do not have to know the reason for the change because they are not capable of judging the best method of doing a job

20. A foreman, by mistake, orders his men to do a job improperly. Of the following, the BEST thing for the foreman to do when he realizes his error is to 20. ...
    A. insist that the job be done as he ordered so that his mistake will not be discovered
    B. admit that he made the mistake and correct the order
    C. tell the men that the order came from "higher up" so that he will not be blamed for the mistake
    D. tell the men that he is merely trying this out to see if it works better

21. The BEST foreman is usually the 21. ...
    A. fastest worker
    B. man who is most familiar with the streets in the area
    C. strongest man
    D. man who is most tactful

22. A good foreman will 22. ...
    A. look after the welfare of his men
    B. demand perfection in the work of his men at all times
    C. make special efforts to impress his superiors
    D. cover up for the actions of his men

23. As a newly appointed foreman, it is MOST important that you 23. ...
    A. show the men who is boss by issuing orders
    B. prove to the men that you know more than they do
    C. become acquainted with the men and their abilities
    D. show the men how friendly you are

24. A foreman who criticizes his department head is a 24. ...
    A. good foreman, because the men will feel he is on their side
    B. poor foreman, because the men will lose respect for him
    C. good foreman, because he will get more work done
    D. poor foreman, because he will have no time to do his own work

25. One of the men in your gang comes to you, the foreman, and complains that the men in the gang have taken a dislike to him and are making trouble for him. Of the following, the BEST thing for you to do is to 25. ...
    A. tell the man he must learn to get along with the other men
    B. report the matter to your superior
    C. call the gang together and tell them they must stop making trouble
    D. investigate the complaint to determine what the problem is

---

KEY (CORRECT ANSWERS)

1.	B	11.	C
2.	A	12.	A
3.	C	13.	B
4.	C	14.	C
5.	B	15.	D
6.	A	16.	D
7.	D	17.	B
8.	A	18.	D
9.	C	19.	A
10.	A	20.	B

21.	D
22.	A
23.	C
24.	B
25.	D

DIRECTIONS FOR THIS SECTION:

Each question or incomplete statement is followed by several suggested answers or completions. Select the one that BEST answers the question or completes the statement. *PRINT THE LETTER OF THE CORRECT ANSWER IN THE SPACE AT THE RIGHT.*

TEST 1

Questions 1-5.

DIRECTIONS: Questions 1 to 5 refer to the following list of supervisory situations. In answering these questions, consider ONLY these situations.

SUPERVISORY SITUATIONS

I. A foreman walks through the shop periodically to check on the work of his workers.

II. A foreman tells his men that if they cannot understand his instructions they should request a transfer.

III. A foreman asks for volunteers whenever a new assignment comes up.

IV. A foreman has periodic meetings with his men to discuss any problems they might have in regard to their assignments.

V. A foreman grants special privileges to those workers who do the best work.

VI. A foreman tells his men that he has been around for a long time and knows more than any of them.

VII. A foreman compliments a worker on his meeting a deadline for a special job. He does this within hearing of the other workers who also worked on this job.

VIII. A foreman is a perfectionist and insists that his men must be perfectionists.

IX. A foreman insists on higher quality work and greater output from the workers who have been longest in the shop.

X. A foreman praises his men at every opportunity but he also criticizes them when he feels it is justified.

XI. A foreman is assigned the responsibility for a new type of equipment. He has trained two of his most competent workers to do this job.

1. The foreman who will probably receive the LEAST amount of respect from his men is the one described in situation 1. ...
   A. I B. II C. X D. XI
2. The foreman whose orders w ll be MOST effective is the one described in situation 2. ...
   A. I B. V C. VI D. VIII
3. The action MOST likely to stir up resentment is the one described in situation 3. ...
   A. III B. IV C. IX D. XI
4. The foreman MOST likely to keep morale high among his men is the one described in situations 4. ...
   A. III and VIII B. VII and XI C. V and IX D. IV and X
5. The foreman whose actions are considered GOOD supervisory practice is the one described in situations 5. ...
   A. I and XI B. II and VIII C. IV and VI D. VII and X

6. The MOST important reason for a foreman to impress upon his men that they should keep their work areas neat and clean is that it 6. ...
   A. makes it easier to find things
   B. increases the available storage space
   C. prevents tools from being damaged
   D. reduces the chance of accidental injury
7. A foreman starts to give instructions to a worker about a certain job and the worker tells this foreman that instructions are not necessary since he knows what to do. This foreman should 7. ...
   A. question the worker as to his intended procedure
   B. let the worker proceed without further instructions
   C. keep the worker under observation while he is doing the job
   D. insist that the worker listen to the full instructions anyway
8. A foreman can develop proper safety habits in his men MORE effectively by 8. ...
   A. posting copies of all accident reports on the bulletin board
   B. letting the men learn from their own mistakes
   C. warning the men that he will penalize anyone who is involved in an accident
   D. observing the unsafe practices of his men and calling for their immediate correction
9. A foreman should impress upon the workers in his group that they should report every unsafe condition which they notice. The ULTIMATE goal of such a reporting system is that each unsafe condition will be 9. ...
   A. examined B. recorded C. corrected D. identified
10. A worker is continuously bypassing his foreman and seeking detailed information about his duties from the foreman's supervisor. From among the following, the MOST probable reason for the worker's bypassing his foreman is that 10. ...
    A. he lacks confidence in his foreman's ability to answer his questions
    B. he is attempting to make trouble for the foreman
    C. the rules and regulations do not prohibit it
    D. the foreman's supervisor is more knowledgeable
11. The policy of a foreman in applying fair and firm discipline in all cases of infractions of the rules, including those of a minor nature, should PRIMARILY be considered a 11. ...
    A. *bad policy,* since minor infractions of the rules do not warrant disciplinary measures
    B. *good policy,* since his subordinates cannot argue that their disciplinary treatment is unjust
    C. *bad policy,* since it will lead to a reduction in the overall efficiency of the organization
    D. *good policy,* since overlooking minor infractions could lead to a more serious erosion of discipline
12. In order to operate efficiently, it is essential that an organization have effective communication from the lower employee levels to the various supervisory levels. This type of communication is BEST obtained 12. ...
    A. through the use of "gripe" sessions with groups of employees

B. by means of formal communication channels between superiors and employees
C. through good day-to-day work relations between superiors and employees in a climate of mutual trust
D. by encouraging all employees to prepare more written reports for review by their superiors

13. In a manual of operations, it is LEAST essential to 13. ...
A. use ordinary vocabulary except where technical language is required
B. describe alternative methods other than those recommended
C. use a loose-leaf system to allow replacement of out-of-date pages of information
D. have conciseness rather than wordiness

14. High motivation of the employees in an organization is more likely to be realized when there are high standards of work and discipline in the organization rather than when there are low standards. Which one of the following answers is MOST in agreement with the preceding statement? 14. ...
A. A foreman who insists on correctness and discipline is not likely to motivate his employees to produce better work.
B. In most cases, there will be a conflict between good human relations and insistence on high quality work.
C. Employees cannot be expected to produce high quality work after they have been criticized severely.
D. Pride in high standards is similar to the pride experienced by a top athletic team.

15. When a new and complicated procedure is being introduced, the LEAST likely reason for a foreman to maintain closer supervision than usual of his workers for the first few weeks is to 15. ...
A. make sure that the men are working
B. make sure that the men understand what they are supposed to do
C. eliminate any unexpected problems that may arise
D. make sure the new procedure is properly coordinated with other existing procedures

16. If a foreman observes one of his workers testing a piece of equipment in an unsatisfactory manner, the MOST acceptable method for the foreman to follow is to 16. ...
A. have another worker instruct the first worker
B. tell the worker how he wants the job done
C. have another worker do the testing properly at a later date
D. take no action until the equipment fails and then point out to the worker why it failed

17. A worker complains to you, the foreman, that his helper cannot be trained to perform a routine job. The FIRST action for you to follow is to 17. ...
A. report to your supervisor that the helper is not competent
B. find out whether the instructor or the learner is responsible for the poor results
C. send the helper back for extra training
D. give the helper a caution

18. As a foreman, if you detect an error in an order issued to you by your supervisor, you should 18. ...
   A. try to carry out the order as issued
   B. correct the error yourself and carry out the corrected order
   C. point out the error to your supervisor
   D. put the order aside until your supervisor detects and corrects the error himself

19. The one of the following which is the MOST important responsibility of a foreman is to 19. ...
   A. train his men in the techniques of giving first aid
   B. decide what qualifications will be required of new men
   C. explain official regulations and policies to his men
   D. determine the length of time required for the various jobs

20. Your supervisor tells you that one of your workers has complained to him about the way you handle the work gang. Since this man has never come to you with complaints, you, as the foreman, should FIRST 20. ...
   A. take the man aside and let him know that you will keep a close watch on him in the future
   B. think over your handling of this man
   C. discuss the situation with the rest of the gang
   D. give an explanation to your supervisor

21. When a worker is being given orders in connection with a specific job, the amount of explanatory information that the foreman should give him depends PRIMARILY on the 21. ...
   A. experience of the subordinate and whether he likes detailed explanations
   B. complexity of the job and how soon it must be completed
   C. experience of the subordinate and the difficulty of the job
   D. amount of interdepartmental coordination involved

22. When criticizing a worker for poor work performance, the BEST course of action for a foreman to take, among the following, is to 22. ...
   A. criticize the subordinate in the presence of others so that the others will also improve their work performance
   B. criticize the subordinate on a personal basis
   C. threaten him with severe disciplinary action
   D. be specific about his criticism and not use generalities

23. Of the following, the procedure which is LEAST desirable in training a newly appointed worker to perform a complicated job is for you, the foreman, to 23. ...
   A. perform each step slowly so that the worker can see how each step is performed
   B. have the worker perform the job under your guidance
   C. demonstrate the different incorrect ways to do the job
   D. describe the job to be done and explain why and how it is to be done

24. In giving verbal instructions to a worker, a foreman should make them direct and concise MAINLY in order to 24. ...
   A. avoid confusing the worker
   B. impress the worker with his efficiency
   C. save time
   D. hold the worker's attention

25. The PRINCIPAL reason for requiring the foreman to submit a supervisory accident report, in addition to the accident report filed by the employee involved, is to 25. ...
   A. accumulate accurate data for statistical purposes
   B. insure that the proper employee is charged with the responsibility
   C. prevent falsification of maintenance records
   D. analyze the cause to help in preventing recurrences

TEST 2

1. At the beginning of each tour it is MOST important for a foreman to 1. ...
   A. check the condition of the plant
   B. complete all of his reports for the day
   C. instruct each worker on the details of his job assignment
   D. review the work load and make work assignments
2. Assume that you have been the foreman in charge of a crew for several weeks when you notice that the men are beginning to show a dislike for you and that this is adversely affecting their productivity. Your BEST procedure is to 2. ...
   A. insist on better discipline in your crew
   B. loosen up on the discipline in your crew
   C. take stock of yourself to determine whether you are to blame
   D. pay no attention to your crew's attitude as foremen are bound to be disliked
3. When analyzing a maintenance department report form to improve its content, the LEAST important consideration is 3. ...
   A. the degree of importance of the report form
   B. the job level of the people using the report form
   C. whether all the material in the report form is up-to-date
   D. the number of years that this report form has been in use
4. When a foreman prepares an accident damage report, it should include 4. ...
   A. a preliminary estimate of repair costs
   B. the names of two witnessed to the accident
   C. a request for an engine piston ring replacement job
   D. a request for an engineering research study of the accident
5. When demonstrating a new procedure to a group of workers, it is ESSENTIAL that a foreman 5. ...
   A. go through the process at the usual speed so that the employees can see the rate at which they should work
   B. show the employees common errors and the consequences of such errors
   C. go slowly and repeat the steps involved at least once
   D. distribute summaries of the procedure during the demonstration and instruct his workers to refer to them afterwards

6. Foremen should set up work schedules to plan for the most effective use of time, men and equipment. Of the following, the MOST important factor for a foreman to consider in setting up a work schedule is the 6. ...
   A. seniority of the men assigned to him
   B. salaries of the men needed for the job
   C. availability of men necessary for the job
   D. standard of performance to be attained by his men
7. One of the MOST important responsibilities of a foreman is to 7. ...
   A. standardize work procedures
   B. redesign equipment which is unsafe
   C. perform the more difficult repair jobs himself
   D. carry out the demands of shop union representatives
8. In order to gain the confidence of his men, a newly-appointed foreman should 8. ...
   A. avoid giving orders until he is completely familiar with his new job
   B. readily admit to his men that he is new at his job and ask for their cooperation
   C. tell his men what was wrong with the way the previous foreman ran the job
   D. threaten to discipline any one who does not agree with his methods
9. A supervisor tells a group of newly-appointed foremen:"You can pass the buck up, but you cannot pass it down." Based on this statement, the MOST accurate statement that would apply when a job is not completed properly is that 9. ...
   A. it is usually possible to find an excuse for every situation
   B. a foreman should be held responsible for the actions of his men
   C. a supervisor should not be held responsible for the actions of his foremen
   D. a foreman should not be held responsible for the work of an inefficient man in his section
10. With regard to the training of workers, it would be CORRECT to say that 10. ...
   A. sound training builds good work habits
   B. a foreman has no time to answer questions on procedures
   C. a worker who makes a mistake will not be able to learn to do a job correctly
   D. newly assigned workers should pick up all of the job knowledge they need by observing the work of others
11. In order to train a new worker, a foreman assigns him to work on a routine job with a more experienced worker. After the customary "breaking in" period, the foreman finds that the new man has made no progress. In this case, the PROPER action for the foreman to take is to 11. ...
   A. report to his superiors that the new man is not interested in learning
   B. extend the training period for as long as it takes to train the man
   C. investigate to determine whether the job is too difficult for a beginner

D. determine whether the experienced worker or the new man is responsible for the poor results

12. In order for a foreman to improve the performance of the men in his section, it is MOST important for him to 12. ...
 A. watch his men constantly
 B. have a set of standards for each job
 C. let each man set his own pace
 D. increase production quotas for all his men

13. A foreman can maintain effective control of his men BEST by 13. ...
 A. refusing to permit changes in established work procedures
 B. liberally interpreting rules and regulations
 C. being firm in important matters
 D. bringing charges against those who disagree with his methods

14. Assume that you are a foreman and your relief appears to be unfit for duty. You should 14. ...
 A. remain with him until you can determine definitely whether he can perform his duties
 B. allow him to take over since the following tour of duty is not your responsibility
 C. order him off agency property, and remain on duty until replaced by another foreman
 D. make an immediate call to your superior

15. A foreman's superior remarks that he could not find the foreman at his post and that his men were idle during the time he was not there. The MOST important thing for this foreman to do as a result of this incident is to 15. ...
 A. improve his supervisory techniques
 B. forget the remark if it is not repeated by this superior
 C. complain to this superior that these men are not dependable
 D. make certain that he is never away from his assigned location again

16. A newly-appointed foreman thinks that the use of group participation in the decision-making process will improve the morale and productivity of his men. For this foreman to make clear to his men the areas in which group participation will NOT be permitted, would be considered 16. ...
 A. *good,* because the men will be unable to take advantage of the foreman
 B. *good,* because certain decisions are beyond the control of both his men and himself
 C. *bad,* because any restraint on the men's freedom of action will dilute the effectiveness of the program
 D. *bad,* because the limitations should be made clear to the group as it becomes involved in a specific area of decision-making

17. A foreman can BEST evaluate the abilities of his men by 17. ...
 A. reviewing their daily attendance records
 B. observing their work on varied types of assignments
 C. finding out what the men think of each other's abilities
 D. asking each worker which methods he thinks would help speed the job

18. Of the following, employee dissatisfactions which are likely to cause the MOST trouble are those which 18. ...
   A. can be relieved only by a change in rules and regulations
   B. smolder for a long time before coming out into the open
   C. deal with a lack of specific tools or test equipment
   D. result from different opinions on how to perform a routine job
19. A foreman who is attempting to help one of his men who has an emotional problem should 19. ...
   A. avoid getting into an argument with the man
   B. tell the man what he should do to solve this problem
   C. accept the face value of the man's statements without probing beneath them
   D. be sure to state his own opinions before hearing lengthy comments from the man
20. One IMPORTANT way a foreman can cooperate with his superiors would be for him to 20. ...
   A. continually push his men to greater production
   B. accept full responsibility for the work of his men
   C. constantly report all details of jobs in progress to his superior
   D. ask for his superior's approval on all decisions made by him (the foreman)
21. In organizing data to be presented in a formal report, a foreman's FIRST step should be to 21. ...
   A. determine the conclusions to be drawn
   B. establish the time sequence of the data
   C. sort and arrange like data into groups
   D. evaluate how consistently the data support the recommendations
22. A foreman is requested by his superior to submit a written report by a certain date. After starting work on his assignment, the foreman realizes he cannot meet the deadline. In this case, it would be BEST for the foreman to 22. ...
   A. immediately inform his superior of the situation
   B. work to the deadline and then ask for an extension
   C. meet the deadline by submitting whatever he has finished at that time
   D. complain to a higher supervisor that the assignment is an unreasonable one
23. When an unusual situation arises and it would take too long to contact a superior for advice, the MOST practical procedure for a foreman to follow is to 23. ...
   A. act according to his best judgment
   B. ask another foreman what he should do
   C. ask his men what he should do
   D. take no action since his superior will resent anything he does
24. New methods of doing work are NOT always accepted by the workers in a shop because 24. ...
   A. routine jobs become boring
   B. workers become used to the old methods
   C. the foreman will have to explain the new methods
   D. a lot of planning is performed before they are introduced

25. If a foreman receives a general bulletin issued by his supervisor criticizing the abuse of the "washing-up" privilege, and his men have NOT been guilty of this offense, he should 25. ...
   A. advise the supervisor that his men are blameless
   B. tell his men about the memo and request their continued cooperation
   C. refrain from mentioning the memo to his men since they may resent it
   D. warn his men that violations of the "washing-up" privilege will not be tolerated

TEST 3

1. As a newly-appointed foreman, you find that maintaining all the records turned over to you by your predecessor is taking too much of your time. Your BEST action is to 1. ...
   A. request an additional clerk to relieve you
   B. stay after regular hours 2 or 3 days a week for this purpose
   C. discontinue keeping the records you believe to be unimportant
   D. consult the supervisor to see whether all the records are necessary
2. Assume that you are called away while you are supervising a job on which several workers and helpers are working. Before leaving this location, the BEST supervisory action for you to take would be to 2. ...
   A. assign each man individually enough work to keep him busy until your return
   B. designate a senior man to be in charge during your absence
   C. leave the job location quietly so the men will not know you have gone
   D. tell the men to take their meal break while you are away and be ready to continue when you return
3. If a foreman observes one of his new workers checking and adjusting a piece of equipment in an incomplete and unsatisfactory manner, the foreman should 3. ...
   A. tell the worker exactly how he wants the work done and see that he does it
   B. call another worker to show the first worker how the work should be done
   C. let him finish the job first and then show him how it should have been done
   D. make a note of this for use when preparing the regular employee performance reports
4. Assume that you, the shop foreman, find out that the supervisor has given one of your men a special job to do without first consulting you about it. As a consequence, the man is not doing the work assigned by you. Your BEST procedure in this case would be to 4. ...
   A. tell the supervisor that going over your head is poor supervision

B. tell the man to complete the special job, and reassign his regular work if necessary
C. tell the man to resume his regular work, setting aside the special job until the supervisor asks about it
D. request an increase in the shop quota to cover future special jobs

5. In looking over your records you notice that a certain worker takes longer to do his assigned work than any other worker who is assigned to similar work. Your BEST initial procedure is to 5. ...
A. consult your superior on the appropriate disciplinary action to be taken
B. give closer supervision to future jobs to which this man is assigned
C. warn the man against deliberate slowness so that he knows you are aware of the situation
D. accept the situation because the work must be averaged out among the faster and slower men in any case

6. It would be necessary for a foreman to obtain approval from the supervisor BEFORE 6. ...
A. personally making a repair on the signal system in an emergency
B. calling the police in an emergency
C. giving one of his workers an oral reprimand
D. arranging with his relief to leave an hour earlier the next day

7. Assume that you are required to transmit an order, with which you do not agree, to your subordinates. In this case, it would be BEST for you to 7. ...
A. ask one of your superiors to transmit the order
B. refuse to transmit an order with which you do not agree
C. transmit the order but be sure to explain that you do not agree with it
D. transmit the order and enforce it to the best of your ability

8. As a foreman, one of the BEST ways of cooperating with your supervisor would be to 8. ...
A. immediately report to him every infraction of the rules which you observe
B. continually report all details of the work under your supervision
C. accept full resppnsibility for the work assigned to you
D. show him how to spread some of his responsibilities among the foremen and workers

9. The one of the following which would be the BEST starting point for an accident reduction and prevention program would be to 9. ...
A. determine the number, kinds, and causes of accidents that have occurred
B. improve working conditions so as to prevent most, if not all, accidents
C. promote an interdepartmental safety and safe practices contest
D. start a "crash" safety program to eliminate all accidents caused by carelesness

10. One of your men has been slightly injured on the job. You have given him first aid. Of the following, the NEXT thing you should do is 10. ...
   A. make out your report
   B. notify his family
   C. take the equipment he was working on out of service
   D. find out the cause of the accident

11. In a program to keep accidents down among your men, it would be LEAST helpful for you to know 11. ...
   A. the man hours lost by your men due to accidents
   B. which men are most often involved in accidents
   C. which types of accidents occur most frequently
   D. why certain accidents occur

12. The MOST effective safety campaign is a campaign which stresses 12. ...
   A. the value of safety to the section or shop
   B. how the loss from accidents is increasing
   C. the safety record of the shop or section
   D. how safe practices protect the worker

13. Suppose that, because of a tight department budget, it has become necessary to reduce service in your section. The one of the following that would probably be the BEST answer to a citizen's complaint about this reduction in service would be 13. ...
   A. a demonstration that the new schedule is adequately meeting the needs of this citizen
   B. an explanation that each citizen should think of the problem as a whole in terms of the needs of the city for all types of services
   C. an explanation that the tight budget made this reduction in service unavoidable
   D. a promise that the former schedules will be restored in the next fiscal year

14. The MAIN reason for overhauling machines on a regular basis is to 14. ...
   A. provide good training for the men
   B. make work for slack times
   C. minimize costly breakdowns
   D. use spare parts before they deteriorate

15. Assume that you are told by another foreman that one of your men violated a safety rule. The BEST action for you, as a foreman, to take is to 15. ...
   A. speak to the man about the incident
   B. tell the other foreman to leave your men alone
   C. watch this man closely in order to catch him next time
   D. give your entire gang a strong talk on safety procedures

16. Of the following, the MAIN purpose of a safety training program is to 16. ...
   A. fix the blame for accidents
   B. describe accidents which have occurred
   C. maintain job progress under unsafe working conditions
   D. make the men aware of the basic causes of accidents

17. The rules and regulations of an agency are usually LEAST useful in 17. ...
    A. encouraging safe practices
    B. relieving the foremen of their supervisory responsibilities
    C. providing a fair basis for any necessary disciplinary action
    D. helping the men in the proper performance of their duties

18. The morale of employees depends to a great extent on whether or not they are able to influence decisions made by management on matters that directly affect the employees. 18. ...
    On the basis of this statement, the one of the following situations which will do the MOST to raise morale is when
    A. the workers have recently received a scheduled pay raise
    B. a suggestion made by a group of workers has resulted in a change in an established procedure
    C. the supervisory staff decides to hold periodic conferences with individual workers to discuss their performance
    D. a department head makes regular visits to job sites to show his interest in each phase of the department's operations

19. Proper training encourages employees to cooperate and lowers operating costs. 19. ...
    The statement MOST consistent with the above statement is that
    A. training is only useful if costs can be lowered
    B. new workers get more benefit out of training than old employees
    C. training is a factor in improving morale and efficiency
    D. employees will refuse to cooperate if they do not receive proper training

20. The BEST of the following reasons for submitting written accident reports as soon as possible after the occurrence of the accident is that 20. ...
    A. a person would tend to include too much detail after a long delay
    B. a person can more accurately recall an event which is fresh in his mind
    C. it is easier to recall events in their proper relationship after a lapse of time
    D. an experienced person can easily recall all essential details after a long delay

21. In order for a foreman to get the work done, his orders to his men must be effective. To make sure that the orders to his men will be most effective, a foreman SHOULD 21. ...
    A. tell his men the reasons why every order must be given
    B. include a great amount of detail in most orders
    C. make the orders very brief in order to keep the attention of the men
    D. give the orders and then make some checks to see if the men understand them

22\. The MOST valid reason why a particular job might have a time limit set on it is that 22. ...
A. the men will be kept continuously busy
B. this particular job is urgent
C. maximum output can be achieved only in this way
D. the best quality of work is thus obtained

23\. A special maintenance procedure is to be used for a new type of machine that has been installed under a recent contract. While this procedure is still new, a close check is to be kept of the maintenance time required. The MOST likely reason for this check is to determine whether the 23. ...
A. special procedure presents any problems
B. work schedules need to be revised
C. special procedure is understood by the workers
D. division should stick with the old procedure

24\. In a large organization, it is CORRECT to say that the many rules, regulations, and procedures tend to 24. ...
A. increase the flexibility of operations
B. encourage employees to do things in a way that they think best
C. reduce the scope of responsibility for a foreman
D. reduce the employees' confusion in carrying out their job assignment

25\. As a foreman, you are inspecting the damage done by water from a broken main leaking into the basement of a store. After inspecting the damage, the owner complains to you about the conduct of the men who made the repair. Of the following, the BEST way of handling this situation is to tell the owner that 25. ...
A. you are there to inspect the damage to the premises only
B. he should make his complaint to higher authorities
C. his complaint will be investigated and, if found correct, proper action will be taken
D. nothing can be done at this time since the men are no longer at this location

TEST 4

1\. Suppose that a foreman makes it a practice to confer with his men on problems which may arise in the section. This practice is 1. ...
A. *desirable*, the men are more familiar with the work than he is
B. *desirable*, valuable suggestions may be obtained
C. *undesirable*, most problems cannot be anticipated
D. *undesirable*, valuable time is wasted

2\. Suppose the Department puts out a regulation that is unpopular with the men. The BEST method for you, the foreman, to use to get your men to obey it is to 2. ...
A. explain the need for it and ask for their cooperation
B. make an example of the first man who violates the regulation
C. publicize the penalty for an infraction
D. tell them you will recommend its repeal after a trial period

3. Suppose there is a regulation that men must notify their sections when they intend to be absent. One of your men stays out without notification. You should, first of all, 3. ...
   A. discuss the matter with your superior officer
   B. find out the reason for his failure to notify the section
   C. find out what other officers have done in similar cases
   D. take no action if the man's absence did not interfere with the work
4. The efficient functioning of any section requires that each man understands what he has to do AND 4. ...
   A. has a voice in setting section policies
   B. is able to assume some of a foreman's duties
   C. knows to whom he is responsible
   D. knows what every other man is doing
5. In general, men will object LEAST to strict regulations if 5. ...
   A. their own foreman is not responsible for them
   B. they affect supervisors as well as men
   C. they are enforced without bias or favor
   D. they have no effect on their wages
6. In general, a man will NOT continue a heated argument if you 6. ...
   A. do not raise your voice in arguing with him
   B. let him have his say
   C. let him know you are above arguing
   D. turn around and walk away
7. A probationary worker under your supervision has not been performing his job satisfactorily. The BEST action for you to take FIRST is to 7. ...
   A. determine whether the cause for his unsatisfactory work can be corrected
   B. find an easier job for him which he can perform satisfactorily
   C. investigate to see if your standards for satisfactory work are too high
   D. recommend that the man be dropped at the end of his probationary period
8. Suppose you suspect that one of your men is loafing on the job. The BEST action for you to take is to 8. ...
   A. do nothing since the man may possibly improve in a short time without action on your part
   B. give him a good lecture on the possible penalties of not doing his work
   C. increase the size of his route so that he'll have no time to loaf
   D. say nothing immediately but carefully inspect his route right after he finishes
9. Suppose a foreman constantly offers his fellow foremen suggestions concerning methods of improving their efficiency. He will, MOST probably, 9. ...
   A. arouse their displeasure and resentment
   B. encourage them to come to him for advice
   C. gain a reputation for wisdom
   D. increase his prestige among them

10. A foreman who sees a man operating a piece of equipment in a faulty manner says to him,"John,you are not operating this machine in the right way," and then walks away. This criticism is defective CHIEFLY because the foreman 10. ...
A. did not suggest a correct method
B. did not pick the best time or place
C. centered his remarks on the person not the act
D. was not sure his judgment was justified

11. Suppose you have unjustly reprimanded one of your men. You SHOULD 11. ...
A. ignore the matter but be more careful in the future
B. make no apology but allow this man's next mistake to pass unnoticed
C. praise the man at the next opportunity even for something small
D. readily admit your mistake to the man you reprimanded

12. Assume that you must criticize one of your men for unsatisfactory work. You SHOULD 12. ...
A. do it only in writing
B. refuse to let the man make excuses
C. soften your remarks with some praise, if possible
D. use stern language to impress him

13. Of the following, the purpose for which a foreman should MOST desire privacy in a conference with a subordinate is to 13. ...
A. answer a question about annual leave
B. give him directions for a new assignment
C. praise him for excellence of his work
D. reprimand him for a mistake he has made

14. Suppose you are in charge of a garage and there are certain unpleasant jobs that must be done. You should assign them to 14. ...
A. all the men in rotation B. the most efficient men
C. the poorest men D. the younger men

15. Suppose that you accidnetally overhear two of your men criticizing your methods of supervision. You should 15. ...
A. ask the men if they want transfers to another section
B. discuss the matter with the men and show them how wrong they are
C. ignore the matter except for purposes of reviewing your methods of supervision
D. watch the men carefully thereafter for signs of poor work

16. Suppose that in checking the adjustments on a machine, you find them faulty or inaccurate. You should 16. ...
A. point out to the operator why the adjustments are wrong and have him reset them
B. reprimand the operator and make the necessary adjustments yourself
C. tell the operator the adjustments are faulty and then walk away
D. tell the operator to get out the instruction manual and read it carefully

17. In breaking in new men on the job, you SHOULD keep in mind that 17. ...
A. men differ in the amount they can learn in a given time
B. men learn more quickly from group instruction than from individual instruction
C. men over twenty learn less quickly than men under twenty
D. the rate of learning will be the same if instruction is the same for all

18. Of the following, a POOR practice for a foreman in charge of a garage to follow in his storeroom is that 18. ..
   A. no material should be issued unless a standard requisition slip is filled out
   B. the material in the storeroom should be so arranged that the oldest is used first
   C. the material which is used most should be located near the entrance of the storeroom
   D. the more an item is used, the less frequently should the supply be checked
19. In face-to-face contact with the public, it is MOST important that a foreman 19. ...
   A. avoid slang and technical terms
   B. turn over complaints to one of his men
   C. have a ready explanation for any complaint
   D. let the citizen break off the conversation
20. In dealing with the public, a foreman should recognize that, in general, people will cooperate with a request when they 20. ...
   A. will get assistance in carrying it out
   B. will learn a new skill as a result
   C. will not be held responsible for the result
   D. understand the reason for the request

KEY (CORRECT ANSWERS)

TEST 1		TEST 2		TEST 3		TEST 4	
1.	B	1.	D	1.	D	1.	B
2.	A	2.	C	2.	B	2.	A
3.	C	3.	D	3.	A	3.	B
4.	D	4.	A	4.	B	4.	C
5.	A	5.	C	5.	B	5.	C
6.	D	6.	C	6.	D	6.	B
7.	A	7.	A	7.	D	7.	A
8.	D	8.	B	8.	C	8.	D
9.	C	9.	B	9.	A	9.	A
10.	A	10.	A	10.	D	10.	A
11.	D	11.	D	11.	A	11.	D
12.	C	12.	B	12.	D	12.	C
13.	B	13.	C	13.	A	13.	D
14.	D	14.	D	14.	C	14.	A
15.	A	15.	A	15.	A	15.	C
16.	B	16.	B	16.	D	16.	A
17.	B	17.	B	17.	B	17.	A
18.	C	18.	B	18.	B	18.	D
19.	C	19.	A	19.	C	19.	A
20.	B	20.	B	20.	B	20.	D
21.	C	21.	C	21.	D		
22.	D	22.	A	22.	B		
23.	C	23.	A	23.	B		
24.	A	24.	B	24.	D		
25.	D	25.	B	25.	C		

CONCISE TEXT

THE FOREMAN

BASIC FUNDAMENTALS OF SUPERVISION AND MANAGEMENT

CONTENTS

I THE FOREMAN

Basic Fundamentals of Supervision and Management

The Job of the Foreman

His duties and responsibilities—His authority—Special problems of the foreman in a small plant.

A foreman is the member of plant management who has been delegated the authority to manage a shop, a function, or a department. He is the one whom the folks in the shop call boss; he is the one they look to for instructions and supervision; and he shoulders the responsibility for all work done by those who report to him. In short, he is the first line of management. In this position he is a key figure both as a production manager and in the plant's relationships with its employees.

His duties and responsibilities

The foreman's job is to use the men, machines, and material assigned to him, for the purpose of getting out production under conditions specified by his superiors. In order to accomplish this objective, he has to perform a multitude of duties and responsibilities.

Human relations duties:

(1) To maintain good relationships with employees, with other departments in the plant, with his superiors, with customers, with unions, and with the public.

(2) To settle grievances—one of his most important duties.

(3) To be familiar with the union contract if there is one, and to run the shop in strict accordance with it.

(4) To be familiar with company policies and interpret them to the workers as management has explained them to him.

(5) To exercise leadership and supervision over the people assigned to him.

(6) To develop and maintain job interest among his employees.

(7) To be available when those who report to him need help or assistance, whether it be of a business or personal nature.

Training duties:

(1) To inform new employees assigned to him about company policies.

(2) To see that new employees receive adequate job instruction.

(3) To train an understudy who can take his place in the event he is absent, or is promoted, or resigns.

Production duties:

(1) *Schedules*—He is expected to get the production out in accordance with schedules prescribed by his superiors; to coordinate the various activities in his department for the purpose of eliminating delays and bottlenecks; and to see that the men, materials, and machines assigned to him are fully utilized.

(2) *Quality*—He is expected to protect the customer from the receipt of faulty products through proper shop precautions.

(3) *Costs*—He is expected to keep production costs within the budget approved by his superiors.

(4) *General production duties*—He is expected to be constantly alert to new methods and procedures which will improve quality, reduce delays, and reduce production costs; he is also expected to see that all production activities conform with plant safety regulations.

Although the duties of a foreman vary by necessity from plant to plant, the ones listed above cover the basic responsibilities which most foremen are expected to shoulder.

His authority

A foreman's authority to make decisions and take action is delegated to him by his superiors. While this, too, must of necessity vary according to the circumstances, most foremen have the following general types of authority.

Supervision.—This involves the authority to exercise full supervision over the men, materials, machines, and supplies assigned to his department—within the limits of company policies.

Use of productive facilities.—This involves the authority to requisition materials, supplies and personnel and incur certain other expenses within the limits of his approved budget.

Maintenance of quality.—This involves the authority to reject any item produced in his shop which fails in his opinion to meet standards of quality as prescribed by his superiors.

Control of jobs.—This involves, for some foremen, the authority to hire and fire.

A foreman's influence is often greater than his specific authority. In certain areas, his superior may regard him as a personal representative, and by virtue of that fact, the foreman's recommendations carry considerable weight. For example, his recommendations concerning personnel, product design, the need for more shop space, new production processes or formulas, and new equipment are given careful consideration. He is usually closer to problems of this nature than any other official in the plant. In the eyes of the workers, the foreman's authority within the shop is all inclusive. If by chance the workers have any question on this score, it is only because the foreman's superiors fail to back him up adequately.

In exercising his authority, and in accepting responsibilities, the foreman acts as a member of a team—the management team. But as in any kind of team work, he must also rely upon his superiors for advice, guidance, and information. Likewise, his superiors are dependent on him for information and suggestions. Crystallizing this concept of teamwork in the minds of all members of management is perhaps the greatest single element of help that can be given to foremen. It encourages greater cooperation between the foreman's shop and other departments, and as a result, the whole company benefits from it. In addition, the foreman *needs* the cooperation and support of every department with which he comes in contact if he is to do his job effectively.

Special problems of the foreman in a small plant

The problems of a foreman in a small plant are often different from those of a foreman in a large company. There are two basic reasons why this is true:

First, small plants can't always afford the services of those specialized departments which are accepted as an essential part of the organization in a large plant. For example, relatively few small plants enjoy the benefit of time-study specialists, of full-time production inspectors, or of separate machine maintenance departments. Although such services may not be handled by specialized personnel, the work still has to be done, and quite often the responsibility for doing it is assigned to the foreman. As a result, the duties and problems of small-plant foremen are usually broader in scope than are those of foremen in a larger plant.

Second, a small plant typically operates on a more intimate basis than a large one. Thus, the relationships between department heads are more personal, and as a result everybody gets to know everybody else much better than is possible in a larger plant. Similarly, department heads are more dependent upon each other for cooperation and help. In such a situation, the foreman has to be doubly alert to the value of good personal relationships, because one unfortunate experience with just one department head in the plant can create a problem between himself and the heads of several other departments. If such an incident occurs, and it results in less cooperation between his shop and other departments, his value to the company is greatly diminished. For this reason the foreman must be an accepted member of the small plant family, in fact as well as theory, or else he has no place in the organization. Therefore, the smaller the plant, the more important personal relationships become.

II

Basic Responsibilities of a Foreman

Getting production out on time—Maintaining quality standards—Holding production costs down—How top management can help foremen improve efficiency.

The foreman has three basic responsibilities:

(1) Getting production out on time.
(2) Getting production to conform with quality standards.
(3) Getting production out at the least possible cost.

Unless a foreman can satisfy these three responsibilities, he is unable to live up to the requirements of his job.

Getting production out on time

The foreman's department has to satisfy certain production quotas because customers want to receive their purchases within a certain period of time. In order to make sure that customers get their shipments within the time specified, one of the foreman's superiors usually develops a production schedule in consultation with the foreman, the sales department, and other interested departments. This schedule then becomes a timetable to which operations must be geared. It places a maximum limit on the amount of time that can be allowed to produce a given number of units. The foreman's objective, insofar as schedules are concerned, should not only be to get production out on schedule, but to get it out as much ahead of schedule as possible, within the resources of his department.

In meeting schedule deadlines, a foreman needs to maintain a set of figures which will tell him at the end of each day how his shop is producing. These figures should tell him that the shop is either adequately or inadequately organized for getting the work out on time. Without such information, his ability to meet schedules is left up to chance.

For example, if a schedule calls for the production of 220 units in a 30-day month, the foreman knows that his shop has 22 work days in which to do the job required. When, during any given day, less than 10 units are produced, he knows he is falling behind schedule and can take remedial action before the production deadline arrives. The foreman's superiors are also interested in these figures and he has a responsibility to keep them posted at all times as to the progress in relation to the scheduled requirements.

Sometimes production falls behind schedule for reasons which have to be referred to higher authority for correction. For example, shipments of raw materials which are needed to produce the finished product may be delayed by the supplier; abnormal sickness may require overtime by others and approval to thus add to expenses; or a machine may break down and delay production to such an extent that the schedule has to be revised.

However, most of the reasons for a shop falling behind schedule are due to internal conditions. Ingenuity of the foreman in identifying the causes and taking remedial action is a test of his ability to hold down the job of foreman. Some of the most common conditions which lie within the authority of the foreman to correct are as follows:

Materials.—The material which goes into the production process may be located in an inconvenient spot. It may be placed where it is so inaccessible or located so far away from the workers that much valuable production time is lost in transporting the material to the machines. Frequently a more efficient arrangement of materials will solve the production lag problem. The foreman has to be alert to the need for having raw materials placed in the best location.

Schedule balance.—Oftentimes the workers engaged in producing a certain part of the product will get ahead of those who are producing other parts. This leaves workers who turn out the finished items standing idle until the parts from the lagging operations shop are available. Since all parts of a given product must be produced on schedule, if the deadline for completed units is to be met, the foreman often has to redistribute the number of men and machines assigned to a given phase of the operation in order to maintain the proper balance of output between the various operations being performed.

Work simplification.—Sometimes the number of operations performed by each worker are so great or so unrelated that the rate of production is slowed down. In such a case production can be speeded up if the job of each worker is broken down so that he has a smaller variety of steps to perform. Frequently, depending upon the nature of the product being produced, this saves time because it reduces the motions in changing from one operation to another. This is a possibility the foreman has to consider, for often it is the key to quicker production.

Machine utilization.—Scheduling the work in proper sequence and with balanced timing is essential if the foreman expects to get maximum use out of each available machine. Regular inspection and maintenance of each machine is also essential if disruptive breakdowns are to be avoided. Down time is expensive: it adds to production costs, to say nothing of the cost of the investment in equipment.

Worker utilization.—Placing workers on those phases of the operation where they can do most effective work is also an aid to meeting production schedules. The foreman has to know how much work each employee is producing, and how his production compares with those around him in order that the slow ones may be singled out for consultation. A worker who is a laggard on one job may do better work on other jobs for which he is better adapted.

In order to meet production requirements, a foreman must maintain a delicate balance between the men, the machines and the materials in his shop. Failure to maintain the balance will result in disrupted production and eventual failure to meet the scheduled deadline.

Maintaining quality standards

Production has to conform with quality standards as prescribed by top management because the customer will return any unit which fails to meet those standards. If such returns reach significant proportions, the reputation and prestige of the plant suffers. Because of the embarrassment arising from the sale of faulty goods to customers, most plants have an arrangement whereby finished goods are inspected before they are shipped to the customer. If, upon inspection, it is found that the product meets all quality standards, it is approved for shipment. If it doesn't, it is rejected and sent back to the shop for more attention. The practice of conducting such inspections is commonly referred to as "quality control." Some plants, especially those which produce critical items (such as parachutes), inspect *every* unit before it is "o. k.'d." But this is expensive, and hence other plants use a type of *statistical quality* control. This is a system whereby only one unit out of a given number of units is checked, with the assumption that if the units checked are found all right, then those in between are likewise all right. The system is becoming more popular in plants and is proving to be effective.

The ideal arrangement is to have finished production inspected by someone who reports to the foreman's superior. This is desirable because the inspector is checking work which the foreman is responsible for producing, and a person who has no responsibility to the foreman, or who has had no part in the production of the items can usually be more objective in his inspections. However, in some of the smaller plants, when inspectors are not carried on the payroll, the inspection function may be delegated to the foreman himself.

Under either arrangement, the foreman should maintain the following information concerning units which are rejected, for use in improving the quality of production:

(1) The number of units rejected within the plant each day, week, or month (whichever is most practicable);

(2) The number of units rejected by customers each day, week, or month;

(3) The name of the worker who produced the unit, if such information is available.

The inspector should keep both the foreman and his superiors advised of the number of units rejected by him. The foreman's superiors should keep the foreman and the inspector advised regarding the number of units rejected by the customers. These records should be used by all three parties to determine whether the number of rejections is excessive. If the number of units rejected exceeds the average rate, it is then up to the foreman to check with the employees responsible and reanalyze his procedures to determine what corrective action is necessary. However, even though the number of rejections is below normal, the foreman still has a responsibility to devise ways and means of effecting further reductions. *The best foremen have the lowest rejection rates.*

Holding production costs down

The use of shop budgets for cost control purposes is the most effective means for making the foreman cost conscious. This is true because he knows that he either has to live within his budget or explain the reasons why. In many plants, the cost accountant prepares a monthly report showing detailed types of expenses as charged to each department in relation to the amount budgeted. This report is useful to top management because it tells how effectively the foreman has controlled his expenses during the previous month. It also serves as a useful supervisory tool for the foreman because it tells him what particular expenditures exceeded the budget, which enables him to know in which area remedial action is necessary.

However, if the foreman is to operate his department as effectively as possible, he can't wait until the end of the month to see how good its performance has been. He usually needs to know on the following day how well he did the day before; in other words, he needs daily expense figures which he can compare with daily budget figures. Such figures on a rounded basis can be easily developed by the foreman himself. All he needs to do is take the previous month's budget report and develop from it average daily budget figures which will give him a fairly good idea of what his maximum daily expenses for the current month should be. He can then develop an estimate on the following day which will tell him what his approximate costs were on the previous day. By comparing the previous day's estimates with his estimated daily expense budget he can get a fairly good idea of how well his expenses are being controlled. If accurate daily performance records can be developed, they of course are more desirable than estimated figures. But many small plants do not have the resources for keeping detailed expense records on a daily basis.

The budget covering the foreman's shop should be prepared by the foreman himself. This not only accentuates the importance of the foreman's responsibility, but it is a more realistic way of preparing a budget. The foreman is closer to the needs of his department than anyone else and if the budget is prepared by someone farther removed from the shop activities, it cannot be all-inclusive and will usually result in the foreman asking for authority to exceed the budget because something was overlooked when it was originally set up. If, after the foreman prepares the budget, his superiors want to cut some of the estimated costs, they then have a more realistic set of figures from which they can estimate what adjusments should be made.

Once a foreman's budget has been established, he should clearly understand that he is expected to keep his expenditures within budgetary limitations. Authority to exceed the budgeted amount should be granted only upon the specific request of the foreman, and only after it has been determined how far he should exceed the budget and for what type of expenditure.

A good foreman not only wants to keep within his budget, he wants to keep his expenditures as far below the budget as possible because he wants to be a good enough administrator to get the production out with the least possible cost.

A budget report to be effectively used as a means of controlling expenditures in the shop, should be issued in final form by the accounting people once a month. It should show both the actual and budgeted monthly expense for the following items:

(1) Direct labor costs.
(2) Indirect labor costs.
(3) Material costs.
(4) Maintenance costs.
(5) Costs of supplies, including tools.
(6) Light, heat, power, and telephone costs.

These are controllable costs (some, of course, are more controllable than others) which the foreman should be conscious of at all times. Other items of a non-controllable nature are also usually included in the budget report. However, as long as the foreman gets reports covering his controllable expenses, he has something which he can use to guide him in his efforts to run a more efficient shop.

The foreman has to be continually conscious of the cost which results from everything that goes on in the shop. Here is a list of some of the major areas on which a foreman has to keep a particularly watchful eye if he is to successfully hold expenses down:

Direct labor costs.—You don't normally cut costs by cutting wages because lower wages attract less efficient people and in the long run, a wage cut often increases labor costs. In fact, some shops have been known to cut costs by raising wages, because through higher wages they were able to attract more efficient people. As a result, less people were needed to turn out the production. The real personnel costs that can and must be controlled are those resulting from absenteeism, turnover, accidents, and improper utilization of workers.

A worker who is guilty of excessive absenteeism increases shop costs because the work schedule of the shop has to be readjusted each time he is absent. That costs money. This is a matter which should be ironed out on a personal basis between the employee and the foreman on the merits of each case.

Excessive turnover is also costly. It is costly in terms of the time and effort used (1) to find and process replacements; (2) to train new employees to your equipment and your methods. Furthermore, it is costly in terms of productivity: it is usually several weeks or longer before the performance of new employees equals that of your experienced workers. The problem of excessive turnover can, in many cases, be clarified by getting from each employee the reasons for his leaving. Both the foreman and the personnel department should be interested in such information. Once the true reasons for turnover are known, corrective action can be taken. High turnover is frequently caused by poor selection, low wages, or poor supervision on the part of the foreman.

When a worker is incapacitated because of an occupational accident, not only does the company temporarily lose his services, but hospital bills, increased workmen's compensation assessments, increased insurance rates, and law suits are apt to result. While not all of these charges would be reflected in the foreman's budget, nevertheless hospital charges and other miscellaneous items become a part of his labor costs.

Failure by the foreman to utilize fully the time an employee is on the job results in increased production costs not only because the worker is being paid for time he has not worked, but because failure to use the regular man-hours available may force the foreman to resort to overtime work in order to get production out on time. A foreman should organize the next day's work before he leaves the shop at night. This is the best way to assure full utilization of all employees.

Indirect labor costs.—These are the labor costs which have nothing to do with actual manufacturing or production costs, such as the cost of labor required to sweep the floors, and repair the machines. In order to keep indirect labor costs to a minimum, the foreman will do well to keep his eye on two major items: the cost of maintaining machines and the cost of handling raw materials. In small plants especially, responsibility for the costs of these functions falls upon the production foreman.

Experience proves that it is cheaper to inspect machines at regular intervals and replace worn parts than to wait until they break down. For when a machine breaks down, its repair usually costs more than keeping the machine in good condition through the process of regular inspection. In addition, a broken-down machine disrupts production, which results in even greater expense.

Perhaps one of the most overlooked expense items in the indirect labor category is the cost of materials handling. It has been estimated, for example, that 36 percent of the labor dollar in the electrical industry is spent in moving materials destined to go into the production process from one place to another. When the material arrives at the plant, it is moved into the warehouse. While there it may be moved several times to make room for other incoming shipments. Then it is moved to the shop, where it may be moved again several times before it is finally used in production. Such a procedure not only makes indirect labor costs excessively high, but also increases the cost of material because the more the material is moved, the more it is in danger of being damaged.

Material utilization.—The foreman can cut the cost of materials by developing standardized procedures for employees to follow in cutting up a piece of material. Most pieces come in standard sizes and lend themselves to formal cutting procedures which can result in considerable savings. A furniture factory, for instance, has found that a piece of plywood which comes in standard 4 x 8 foot sizes can make a small bookcase with only one-half of a board foot to spare, if the cutting is done in a certain way. If the piece is cut any other way, it takes two such pieces, with a lot of material left over. Most of it cannot be used at all, and the part that is usable can be used only by disrupting the assembly line. Material costs are a big factor in production and proper utilization can help hold production costs down.

Machine utilization.—Machines cost money whether they are utilized or not because they depreciate and become obsolete. To justify these costs, machines should be fully utilized at all times through proper work organization. Therefore, before the foreman goes home at night, he should know exactly how every employee and machine will be utilized when the shop opens up the next day.

Methods improvement.—To do his job of keeping costs down to a minimum, a foreman must be, in addition to everything else, an "efficiency expert." He must constantly be on the lookout for better, cheaper, and easier ways of getting the production out. He should consider such things as the number of steps a worker must take in going from one operation to another; a rearrangement of equipment will save money if that will cut down on the number of steps required. He needs to consider the number of motions a worker must go through to complete a given operation; if a change in procedure will make some motions unnecessary, he has saved the company money.

In attempting to make his shop as efficient as possible, the foreman can make progress if he will look to his employees for ideas. They are in a position to see many things which he is either unable or too busy to see. Some of the best ideas for increasing efficiency in plants come from employees. A formalized suggestion plan promoted by the foremen will serve as a stimulus for employees to develop new ideas and turn them in. A suggestion plan if properly promoted can be one of the most effective steps toward greater plant efficiency.

How top management can help foremen improve efficiency

There are at least three things which top management can do which will help the foreman improve the efficiency of his shop:

Let him have a voice in the purchase of materials, tools and equipment.—The foreman lives every day with the materials, tools and equipment in his department. He knows the good and bad points of every brand-name in his shop. He knows which items speed up production and which ones do not. While there is no need to give the foreman the last word on purchases made for his shop, he should be given the opportunity to express his opinions on the matter. In most cases, he will be able to offer ideas that will influence the type of purchases to be made. This should result in having more efficient equipment and in improved morale of the work force.

Let him select his own employees.—Recruiting of personnel should be done through the personnel office, based upon qualifications submitted by the foreman. But the foreman should have the authority to determine which of the applicants submitted by the personnel office he wants to have in his department. This enables the foreman to select those people whom he feels will be most capable of doing the job the way he wants it done. Such an arrangement is conducive to a more harmonious shop, and helps give the foreman the prestige necessary to carry out his responsibility. The personnel office should continue to make personnel policies to which the foreman should of course conform, but a foreman should have the freedom to select his own employees within the limits of those policies.

Give the foreman authority to recommend or deny wage increases.—If the foreman wants to have the wages of one of his employees raised, he should have the authority to recommend the increase. Moreover, every effort should be made to grant it, provided it does not conflict with the plant's overall personnel policies. This procedure helps also to increase the prestige of the foreman and gives the workers assurance that their ability to get wage increases is not dependent solely upon some higher authority who is unfamiliar with their work. Conversely, if the worker's performance does not, in the opinion of the foreman, justify a proposed wage increase, he should be authorized to deny it. If the plant is a union shop, of course, suitable adjustments in these policies would have to be made.

III

The Foreman's Training Functions

Training new employees—Training employees promoted to new jobs—Training an understudy—Other training responsibilities.

A foreman has to assume certain training functions—whether the plant has a training department or not—because there are certain procedures and techniques which are peculiar to many jobs. In a small plant, it is often difficult for anyone outside of the department in question to be entirely familiar with its procedures and techniques. Moreover, procedural changes often take place so fast that a centralized training department would find it almost impossible to keep on a current basis.

Training new employees

The foreman has a threefold training responsibility for new employees:

Explaining company policies.—He has a responsibility, at the time the new employee reports for work, to brief him on all company policies which directly relate to him individually; for example: regulations regarding pay, including lost time and overtime, seniority provisions, seniority increases, union contract provisions, and promotions; company-sponsored insurance plans; workmen's compensation, and the rights of the employee in connection with each. These are items of such importance to the new employee that they are an essential part of his orientation.

Instruction covering shop operations.—The foreman also has a responsibility to introduce the new employee around the shop, explain the functions of each worker, point out the different types of equipment and what they are used for, and explain shop safety regulations. He also should explain the relationship of the new employee's job to the rest of the department and to the rest of the company, and briefly outline the organization of the company as a whole. This helps the employee to get a proper perspective of his new job and the plant.

Coaching covering the job of a new employee.—The foreman has a responsibility to see that the new employee is properly broken in on his new job. This is usually done by assigning him to an experienced employee who provides detailed instruction and assistance until the new worker is qualified to handle his assignments alone.

When a new employee is hired for a complicated job requiring several different types of operations, foremen frequently break the job down into a series of simple steps, to facilitate training. When the new employee learns how to handle the first operation, he is then instructed on how to do the second operation until he has mastered that step, and so on. This is known as the "job dilution" type of training. It was widely used in shops both large and small during World War II. This method of breaking in a new employee on a complicated job usually accelerates the training period with a minimum of effort.

Training employees promoted to new jobs

When an employee is promoted or transferred to a different job, the foreman has a responsibility to see that adequate instruction is given the worker on his new job until he is qualified to handle his assignments alone. As in the case of a newly hired employee, such training usually requires that an experienced employee provide the necessary detailed instruction.

Training an understudy

The foreman has a responsibility to develop an understudy who can take over his job if and when he is promoted or leaves the company.

Some men hesitate to develop an understudy because they feel he may eventually prove to be a competitor for the job of foreman. Yet a foreman owes it to himself to have an understudy who can take his place. Failure to have one may prevent him from being promoted to a better job, simply because there is no one to take his place. In many cases, an oversight of this nature has stood in the way of a foreman's promotion.

From the plant manager's viewpoint, an understudy is desirable because of the possibility that the foreman may get sick, leave the company, or be eligible for transfer or promotion.

Other training responsibilities

Foremen are often given other types of training responsibilities, among which are the following:

Orientation of salesmen and other company personnel.—Foremen are often called upon to explain the production process to salesmen for their information in selling the product to customers. Others in the company who hold management positions often call upon the foreman to give them the same briefing so they can get a more complete picture of plant operations. Sometimes foremen are called upon to explain the production process to customers who want to know how the product they buy is made.

Special training sessions for employees.—When major changes are made in the production process; plant mechanization is increased; wage payment systems are changed; time-study plans are introduced; plant operations are reorganized, or special sales and advertising campaigns are initiated, the foreman has a responsibility to explain to his employees the reasons why the changes are taking place, and how these changes will affect them.

Some plants, realizing that the foreman is the company so far as his employees are concerned, rely upon him to hold employee training sessions in cooperation with other supervisors for the purpose of acquainting employees with the operations of the company in general, why it operates the way it does, what the plant balance sheet shows, what the current business conditions are and how they affect the operations of the company, and what the prospects of the company are for the future. Such sessions help to educate the worker with respect to economics and the free enterprise system. Most research organizations agree that employees are in need of such training because many of them are often ill-informed on matters of this nature.

IV

The Foreman's Personal Relationships and Contacts

Relationships with workers—Contacts with supervisors in other departments—Relationships with superiors—Contacts with unions and union officials—Contacts with the public.

In the average small plant, each function and each department is so interdependent that the foreman who succeeds in establishing good personal relationships throughout the company has a much easier time performing his job. Since the foreman's shop is dependent upon other departments for materials, equipment, supplies, and personnel, the amount of cooperation he is able to get from other departments has a definite effect upon the performance of his shop.

Relationships with workers

The value of good personal relationships between the foreman and his employees cannot be overestimated, because it is through such good relationships that he is able to settle more grievances and avert more union complaints than any other member of the management team. A foreman in one small plant, for example, is known to have settled an average of one grievance a month for the past 5 years. His technique is simple. He listens until the worker gets the complaint off his chest. Then, if the worker is upset at the time the complaint is made, he waits until the next day, when the subject can be discussed more objectively. If after investigation, he finds that the complaint is valid, he takes corrective action—provided his department is at fault. If another department is at fault, such as the payroll or personnel office, he requests the department in question to take corrective action. And he follows up on the matter until the case is closed.

If a foreman does not enjoy good personal relationships with his employees, they may not choose to discuss their grievances with him. They may, instead, take them direct to the union and thereby strain the relationship between union and plant before the matter is settled. There is evidence that good foreman-employee relationships have been responsible for smoothing out many grievances which otherwise would have resulted in strikes.

Foremen who enjoy the best personal relationships with their employees are those whose actions prove that they sincerely have the interest of their workers at heart. Such action becomes evident in the foreman's efforts to provide his employees with adequate equipment and supplies; when he tries to make the job easier for them; when he tries to improve their comfort while on the job; when he takes an interest in seeing that they get their pay checks on time and in the right amount; when he makes allowances for illness in the family or other personal difficulties of his workers; and when he tries to promote harmony among all of the people in his department.

Many foremen find that occasional parties after work help to improve morale by enabling everybody to get acquainted. Also, occasional social visits to the homes of his employees improves their job interest and enables the foreman to know them better. If a foreman operates his shop on a strictly impartial basis, visits to the homes of his employees are a great help in improving morale.

Contacts with supervisors in other departments

Much that is used within a shop, such as materials and supplies, comes from other departments. The foreman therefore has to have the cooperation of these other departments if he is to get from them the kind of service he needs in order to keep his shop running efficiently. While it is true that he can appeal to his superior if the cooperation he wants from other departments is not adequate, this procedure takes time. It creates even worse feelings, and in the long run, the foreman who made the complaint loses more than he gains. There is no substitute for good, spontaneous, interdepartmental cooperation.

Relationships with superiors

The foreman has a responsibility to keep his superiors informed regarding his problems, his needs and the overall operations of his department. Top management relies on information of this nature for control and planning purposes. He also has a responsibility to keep his superiors currently informed regarding those problems and situations affecting the performance of his shop which are outside the scope of his authority to handle. Failure to keep his superiors properly informed may result in eventual embarrassment to the foreman because a small problem, over which he has no control, may later develop into a major problem if it is allowed to go unattended.

Contacts with unions and union officials

A foreman's contacts with unions and union officials must at all times be honest and sincere. There is no other way to deal effectively with unions. It is a foreman's responsibility to respect and abide by the agreement which his company and his employees' union have developed. The foreman helps to promote better management-labor relations when the union knows that he fully respects and makes a sincere effort to support all provisions of the agreement which his company has signed.

Contacts with the public

The foreman is often called upon to handle plant tours for groups who want to see how products are manufactured. This places a public relations responsibility upon the foreman to give such visitors a good impression of the plant. Industrial management is becoming more and more aware of the need to acquaint the general public with the facts about its plants and the contributions which they make to the community. One of the more effective ways to do this is through plant tours.

The heads of many plants encourage the foremen and other supervisors to take part as much as possible in civic activities designed to improve the community in which the plant is located. They take the position that since the plant is a part of the community, the leadership in that plant should take a definite interest in community affairs. When the foreman engages in such activities, he not only advertises his company as being a civic-minded organization, but he assumes community responsibilities which contribute to the growth of both himself and the community.

V

Developing Better Foremen

Qualifications which foremen need—Finding prospects—What kind of training?—Solving the problem in small plants—Measuring the results—How top management promotes foreman development.

The problem of developing better foremen is more than just a matter of putting them through a foremanship training course. In fact, a training program is only one of several factors which have a bearing on the development of foremen. Before there can be effective training, it must first be determined that the foremen and prospective foremen possess the basic qualifications which such a job demands. Any time spent in training people in foremanship is wasted unless those being trained actually have the qualifications necessary to absorb and use such training effectively.

Even then, the maximum effectiveness of a foreman training program cannot be realized unless top management follows up such training with an arrangement which provides proper recognition of the role which a foreman is expected to play in his organization. If a foreman completes a training course which stresses the importance of his job (and virtually all foremen training courses do stress this) and later finds that the importance of that job is minimized through lack of proper authority and recognition, part of the value of the training course is lost.

The problem of developing better foremen is, therefore, one which involves three basic considerations: (1) the prospective foreman must be qualified for the job, (2) special training for development on the job should be available, and (3) the importance of the role the foreman is expected to play in the operation of the plant should be fully recognized.

Qualifications which foremen need

The wide publicity given in recent years to the importance of the role played by foremen in industry has developed a realization by top management that, in the eyes of the worker, the foreman is the company, and, in the eyes of the community, the company is judged in part by the impression the foreman makes on the general public. This realization has resulted in a more careful selection of people who are placed in such jobs. Since the foreman is charged with the responsibility of getting the job done in the quickest possible time and with the least possible expense, he must be a highly qualified person. Recent studies by management associations and research organizations generally indicate that top management today is looking for the following nine qualifications in the people who are considered for the position of foreman:

1. *Leadership ability.*—A recent survey taken among 300 companies indicates that leadership ability is the first qualification that top management looks for in a prospective foreman. The modern concept of a foreman is one who provides leadership, not one who simply gives orders. His ability to get production out on time with the least possible cost is dependent to a large extent upon how effectively his leadership can stimulate job interest and satisfaction on the part of his employees. A foreman is a good leader when his employees turn out a good job not because they have to but because they as employees want to do a good job.

2. *Organizing ability.*—A foreman has men, machines, and materials assigned to him. He must have the ability to organize all three factors in an orderly, clear-cut, and simplified manner, whenever circumstances demand it. Nothing will break down the morale of a shop quicker than a situation where the work is so poorly organized that the employees are, figuratively, stepping on each other, or where some employees are overworked while others are standing around waiting for work. Unless a foreman has the administrative skill necessary to organize the work of his shop in a manner that will hold bottlenecks and trouble spots to a minimum, his other qualifications as a foreman will be of little value.

3. *Character.*—A foreman has to be honest and fair with his employees at all times. In this respect his reputation must be above reproach. One of the costliest operators a company can have is a foreman or supervisor who arouses suspicion and distrust. Employees will invariably produce less work under such supervision. Their morale will not only be low, but they will spend valuable production time worrying about what the foreman is going to do next. This results in rumors and secret discussions between employees on company time, which increases the time and the cost required to get out production.

4. *Judgment.*—Employees do better work when they have confidence in the foreman's judgment. Especially is this true in a shop where poor judgment could result in accidents to employees. Very often employees are asked to do a job or follow a procedure in which they do not see any merit. Confidence in the judgment of the foreman will tend to allay any fears that the foreman doesn't know what he is doing.

5. *Technical skill and mechanical ability.*—The *ideal* foreman should be able to run any machine in the shop as well if not better than any employee under his supervision. It helps strengthen his position as a leader. However, this is not always possible because some machines cannot be operated at peak efficiency unless a person has daily experience with them. Nevertheless, a foreman should know the capabilities and limitations of every machine in his department and should be able to tell new people how to operate them effectively.

6. *Education.*—While it is true that more and more foremen have had college training, a college education is not regarded as a necessary qualification. Some of the country's best foremen have never been to college. Insofar as education is concerned, top management is looking, first, for foremen whose background provides a broad commonsense viewpoint and practical understanding. Although college training helps in the development of these qualities, it is also recognized that such qualities can be the outgrowth of past experience.

7. *Initiative.*—A foreman must be a self-starter. If he is to discharge his responsibility as a manager of a function, he must be capable of taking action necessary to keep production going with only occasional supervision from his superiors. Any foreman who requires constant supervision is not performing the functions of a foreman.

8. *Human interest.*—A foreman should be interested in all things which affect the well-being of his employees. This involves personal as well as shop problems because personal problems frequently affect performance on the job. Moreover, the foreman is, oftentimes, the only person to whom employees can go for counsel and advice on problems of this nature. When a worker knows that the foreman has his personal interest at heart, his morale and job interest are greatly stimulated.

9. *Physical and mental requirements.*—It is needless to say that a foreman must have the physical and emotional capacity to carry his responsibilities and perform his duties with the vigor which the job requires.

Finding prospects

Companies generally try to find people *within their own organization* who can be promoted to the position of foreman whenever a vacancy occurs. They do this because it has a definite bearing on the growth and development of foremen into better supervisors. When a man is promoted from the ranks to the position of foreman, he is given evidence that the company is trying to follow the policy of "promotion from within." This gives him the feeling that if he does a good job as foreman he may have a chance at some later date to be considered for an even better job. A foreman who feels he can look to his own company for advancement, rather than having to look to a competitor, is able for psychological reasons, to give more of himself to his job and his company. This procedure also helps stimulate job interest on the part of the foreman's employees because they, too, feel that if they do a good job where they are, they may later be advanced to the position of foreman. More and more employees are looking for jobs that offer career opportunities, and as long as a company holds to the policy of promotion from within, workers can feel that their present company offers career possibilities.

What kind of training?

Programs for training foremen take various forms. Among them are: programs designed to prepare candidates for promotion to the job of a foreman; programs designed to stimulate the development of existing foremen so they can do a better job; refresher courses for foremen who have already been trained; and programs designed to train foremen for higher positions in the organization.

The fact that foreman training is so strongly emphasized by industry does not mean that the foreman of today is lacking in capability. On the contrary, most foremen are capable, well-qualified people. However, attitudes of people are constantly changing, new labor relations philosophies are being developed, and new processes and new techniques are being introduced. It is only good business to expose foremen to training sessions which will give them the information necessary to keep abreast of the changing industrial picture. The primary reason why such careful attention has been given to the subject of foremen training in recent years is the long-unfilled need which resulted from the time when it was not felt necessary to train foremen at all. It was felt that this need could be adequately met through supervision of the foreman by his superiors. Here is another example of the fact that industrial philosophies and practices change.

A foreman on the job today already has a certain amount of leadership ability, organizational ability, understanding of people, technical skill, and mechanical ability. These are qualifications he has to possess if he is to hold down his job. He can't get production out on time without them. The question is: Can the foreman's ability to lead and understand people, and organize his shop, be improved with training designed to help him in these areas? The answer is: "Yes." Leaders from both large and small companies agree that training in foremanship can definitely help the foreman to develop his abilities and thereby enable him to do a better job in the plant. In addition, after a foreman has taken a course in leadership training, he needs refresher courses from time to time because under pressure of meeting schedules and deadlines even the most conscientious supervisor forgets or overlooks part of what he already knows. Experience proves that the art of getting things done through people is a subject which supervisors can, with profit, be coached in—*at regular intervals.*

The foreman also needs training in shop organization and in work simplification. There are certain basic principles of good organization which can be applied to most any department. Also, new techniques are developed from time to time which permit plant procedures to be simplified with a saving of time and money. Any foreman training program should deal with the subject of shop organization techniques because good organization reduces waste and production costs and improves employee morale.

Solving the problem in small plants

Because of practical considerations, small plants are limited in their ability to develop and promote their own formal training programs for foremen. Nevertheless, many of them consider the need for this training to be of such importance that they are discovering ways and means for making it available in spite of limited resources. In general, small plants are working through one of four general methods.

Cooperation with other small plants.—In areas where a number of small plants are located, arrangements can often be made whereby several plants jointly hire a training specialist who develops a foreman training program designed to fit the particular needs of the cooperating plants. Such projects are financed by the cooperating firms on a joint basis. Arrangements of this type have proved successful in Chicago, Pittsburgh, and Philadelphia. The idea was developed with the help of management consulting firms.

Cooperation with colleges and universities.—Some colleges and universities, in cooperation with small plants, have developed foreman and supervisory training programs designed to meet the needs of plants in their particular areas. These programs are usually offered in connection with adult training courses given by the colleges. Those who attend usually get the benefit of instruction both from college professors who have specialized in training problems and from industrial experts. While there are relatively few such arrangements—at least so far as foremen are concerned—plants that have participated in projects of this type regard them as most successful. An alternative method of training foremen is the home study course. Subjects taught by accredited home study schools include quality control, report writing, safety engineering, and others.

Use of programs developed by larger companies.—Some small plants purchase foreman training programs from larger companies which are in the same general type of business. They use such programs to conduct training sessions under the leadership of someone on their own staff. Some large companies are very generous in letting other companies use their material. While this arrangement has its disadvantages, it has been known to produce desirable results, especially where the directors of such training sessions are skilled in conference leadership techniques. Those who use this arrangement point out that, in the final analysis, foremen will train each other if (1) they have an adequate text upon which to base their discussions, and (2) the leader of the training sessions is gifted in stimulating group discussions in connection with problems which foremen as individuals face on their particular jobs.

Where a small company has a close relationship with one of the larger companies, arrangements can often be made whereby the small plant sends a member of its staff to attend the foreman training course of the large company. That person ultimately returns and serves as an instructor for foremen in his own plant. It should be noted, however, that training programs developed for large companies need to be modified if they are used in small plants.

Use of supervisory programs developed by management associations.—Some small plants solve their foreman training problem by making use of the staff facilities of national management associations. The assistance rendered by such organizations to small plants has produced excellent results. Among those associations which have taken an active interest in foremen training are the National Association of Manufacturers, some local Chambers of Commerce, the National Association of Foremen, the Foremen's League, and the American Management Association. These groups offer a variety of services which can and are being utilized by small companies as follows:

(1) Some have training specialists who are available to small businesses for the purpose of conducting foremen and supervisory training programs.

(2) Some have training specialists available as consultants for the purpose of helping small establishments set up their own training program.

(3) Some have foreman training programs of various types which small plants can use in conducting their own training sessions.

(4) Some offer a generalized type of foreman training through the use of foremen's clubs. Such clubs are designed to promote a better understanding by the foreman of his responsibilities as a member of management.

Measuring the results

Since a training program for foremen deals with leadership and organizational problems, the results of the training can usually be measured only in terms of the foreman's performance on the job. There is no way to predict from classroom discussions how valuable the training will be to the company. The results of such training, if they are to be of value, must be reflected in the performance records of the foreman's department. If the course has been a success, proof will be found in increased ability to get work out in less time, with less expense, and with fewer employee grievances than previously. While it may take weeks for any results to become apparent, a good foreman training program will help to accomplish this objective. Following are the specific items by which the foreman's performance can be measured to determine what results have been obtained: overtime expense, number of production units which have been rejected, amount of material wasted, number of workers required to get the work out, number of hours the machines are utilized, the number and nature of employee grievances which are filed, and the amount of absenteeism. If the training program has been successful, there should in due time be an improvement in each of the categories listed above.

How top management promotes foreman development

A training program alone is not sufficient to encourage the development of foremen to the maximum extent. After a foreman has completed his training, he wants to see evidence that his job is as important as the training program led him to believe. If he is to be regarded as a part of management, he wants to be given the authority, prestige, and recognition which goes with it. Once he has been given proper recognition, this tends to serve as a constant reminder of the importance of his job and encourages him to make greater efforts toward self-improvement on his own initiative.

There are several ways in which top management can provide the proper recognition to foremen:

Incentive payments.—Many firms have the policy of paying bonuses to their top management in proportion to their salaries, based upon how much profit the company makes each year. By spreading such bonus payments to all levels of management, foremen and other supervisors are made to feel that they really are a part of management and that the more efficiently they run their shops, the greater the chance that their annual bonuses will be increased.

Authority.—It is not enough to spell out the duties and responsibilities of the foreman. He must be given suitable authority. Top management should also make it clear that they will back up the foreman in any decisions that he makes within the scope of his duties and responsibilities. This helps to promote the growth of the foreman because it keeps him alert to the fact that he is, in reality as well as title, a part of management.

Prestige.—Top management can develop a sense of responsibility and job satisfaction in its foremen, by use of the following other devices:

(1) A bulletin from the president to all levels of management informing them of any major changes in company policy. It lets the foreman and other supervisors know that the top management prefers for supervisors to get the story direct rather than through the grapevine or the newspapers.

(2) Staff sessions called by top management of a small plant for the purpose of getting advice on company problems have proved effective in developing a broader sense of responsibility among foremen.

(3) Occasional banquets where all levels of management are invited for getting-acquainted purposes helps to stimulate job enthusiasm on the part of foremen.

A foreman who is properly trained and who enjoys proper recognition of his status is one of the most important people in industry today. The extent to which he develops his abilities to do the job expected of him is, in large measure, dependent upon how much interest top management in the small plant is willing to take in him.

ANSWER SHEET

EST NO. ______________ PART ______ TITLE OF POSITION ______________________________
(AS GIVEN IN EXAMINATION ANNOUNCEMENT - INCLUDE OPTION, IF ANY)

PLACE OF EXAMINATION ______________________________ (CITY OR TOWN) ______________ (STATE) DATE ______________

RATING

USE THE SPECIAL PENCIL. MAKE GLOSSY BLACK MARKS.

1	A	B	C	D	E	26	A	B	C	D	E	51	A	B	C	D	E	76	A	B	C	D	E	101	A	B	C	D	E
2	A	B	C	D	E	27	A	B	C	D	E	52	A	B	C	D	E	77	A	B	C	D	E	102	A	B	C	D	E
3	A	B	C	D	E	28	A	B	C	D	E	53	A	B	C	D	E	78	A	B	C	D	E	103	A	B	C	D	E
4	A	B	C	D	E	29	A	B	C	D	E	54	A	B	C	D	E	79	A	B	C	D	E	104	A	B	C	D	E
5	A	B	C	D	E	30	A	B	C	D	E	55	A	B	C	D	E	80	A	B	C	D	E	105	A	B	C	D	E
6	A	B	C	D	E	31	A	B	C	D	E	56	A	B	C	D	E	81	A	B	C	D	E	106	A	B	C	D	E
7	A	B	C	D	E	32	A	B	C	D	E	57	A	B	C	D	E	82	A	B	C	D	E	107	A	B	C	D	E
8	A	B	C	D	E	33	A	B	C	D	E	58	A	B	C	D	E	83	A	B	C	D	E	108	A	B	C	D	E
9	A	B	C	D	E	34	A	B	C	D	E	59	A	B	C	D	E	84	A	B	C	D	E	109	A	B	C	D	E
10	A	B	C	D	E	35	A	B	C	D	E	60	A	B	C	D	E	85	A	B	C	D	E	110	A	B	C	D	E

Make only ONE mark for each answer. Additional and stray marks may be counted as mistakes. In making corrections, erase errors COMPLETELY.

11	A	B	C	D	E	36	A	B	C	D	E	61	A	B	C	D	E	86	A	B	C	D	E	111	A	B	C	D	E
12	A	B	C	D	E	37	A	B	C	D	E	62	A	B	C	D	E	87	A	B	C	D	E	112	A	B	C	D	E
13	A	B	C	D	E	38	A	B	C	D	E	63	A	B	C	D	E	88	A	B	C	D	E	113	A	B	C	D	E
14	A	B	C	D	E	39	A	B	C	D	E	64	A	B	C	D	E	89	A	B	C	D	E	114	A	B	C	D	E
15	A	B	C	D	E	40	A	B	C	D	E	65	A	B	C	D	E	90	A	B	C	D	E	115	A	B	C	D	E
16	A	B	C	D	E	41	A	B	C	D	E	66	A	B	C	D	E	91	A	B	C	D	E	116	A	B	C	D	E
17	A	B	C	D	E	42	A	B	C	D	E	67	A	B	C	D	E	92	A	B	C	D	E	117	A	B	C	D	E
18	A	B	C	D	E	43	A	B	C	D	E	68	A	B	C	D	E	93	A	B	C	D	E	118	A	B	C	D	E
19	A	B	C	D	E	44	A	B	C	D	E	69	A	B	C	D	E	94	A	B	C	D	E	119	A	B	C	D	E
20	A	B	C	D	E	45	A	B	C	D	E	70	A	B	C	D	E	95	A	B	C	D	E	120	A	B	C	D	E
21	A	B	C	D	E	46	A	B	C	D	E	71	A	B	C	D	E	96	A	B	C	D	E	121	A	B	C	D	E
22	A	B	C	D	E	47	A	B	C	D	E	72	A	B	C	D	E	97	A	B	C	D	E	122	A	B	C	D	E
23	A	B	C	D	E	48	A	B	C	D	E	73	A	B	C	D	E	98	A	B	C	D	E	123	A	B	C	D	E
24	A	B	C	D	E	49	A	B	C	D	E	74	A	B	C	D	E	99	A	B	C	D	E	124	A	B	C	D	E
25	A	B	C	D	E	50	A	B	C	D	E	75	A	B	C	D	E	100	A	B	C	D	E	125	A	B	C	D	E

ANSWER SHEET

TEST NO. ____________ PART ______ TITLE OF POSITION ____________________
(AS GIVEN IN EXAMINATION ANNOUNCEMENT - INCLUDE OPTION, IF ANY)

PLACE OF EXAMINATION ____________________ (CITY OR TOWN) ____________________ (STATE) DATE ______

RATING

USE THE SPECIAL PENCIL. MAKE GLOSSY BLACK MARKS.

1 A B C D E	26 A B C D E	51 A B C D E	76 A B C D E	101 A B C D E
2 A B C D E	27 A B C D E	52 A B C D E	77 A B C D E	102 A B C D E
3 A B C D E	28 A B C D E	53 A B C D E	78 A B C D E	103 A B C D E
4 A B C D E	29 A B C D E	54 A B C D E	79 A B C D E	104 A B C D E
5 A B C D E	30 A B C D E	55 A B C D E	80 A B C D E	105 A B C D E
6 A B C D E	31 A B C D E	56 A B C D E	81 A B C D E	106 A B C D E
7 A B C D E	32 A B C D E	57 A B C D E	82 A B C D E	107 A B C D E
8 A B C D E	33 A B C D E	58 A B C D E	83 A B C D E	108 A B C D E
9 A B C D E	34 A B C D E	59 A B C D E	84 A B C D E	109 A B C D E
10 A B C D E	35 A B C D E	60 A B C D E	85 A B C D E	110 A B C D E

Make only ONE mark for each answer. Additional and stray marks may be counted as mistakes. In making corrections, erase errors COMPLETELY.

11 A B C D E	36 A B C D E	61 A B C D E	86 A B C D E	111 A B C D E
12 A B C D E	37 A B C D E	62 A B C D E	87 A B C D E	112 A B C D E
13 A B C D E	38 A B C D E	63 A B C D E	88 A B C D E	113 A B C D E
14 A B C D E	39 A B C D E	64 A B C D E	89 A B C D E	114 A B C D E
15 A B C D E	40 A B C D E	65 A B C D E	90 A B C D E	115 A B C D E
16 A B C D E	41 A B C D E	66 A B C D E	91 A B C D E	116 A B C D E
17 A B C D E	42 A B C D E	67 A B C D E	92 A B C D E	117 A B C D E
18 A B C D E	43 A B C D E	68 A B C D E	93 A B C D E	118 A B C D E
19 A B C D E	44 A B C D E	69 A B C D E	94 A B C D E	119 A B C D E
20 A B C D E	45 A B C D E	70 A B C D E	95 A B C D E	120 A B C D E
21 A B C D E	46 A B C D E	71 A B C D E	96 A B C D E	121 A B C D E
22 A B C D E	47 A B C D E	72 A B C D E	97 A B C D E	122 A B C D E
23 A B C D E	48 A B C D E	73 A B C D E	98 A B C D E	123 A B C D E
24 A B C D E	49 A B C D E	74 A B C D E	99 A B C D E	124 A B C D E
25 A B C D E	50 A B C D E	75 A B C D E	100 A B C D E	125 A B C D E